The Untold Power of the Marriage of Art and Technology

By John M. Eger

The Untold Power of the Marriage of Art and Technology

By John M. Eger

First published in 2022
as part of the *Technology, Knowledge & Society* Book Imprint
doi: 10.18848/978-0-949313-04-1/CGP (Full Book)

Common Ground Research Networks
2001 South First Street, Suite 202
University of Illinois Research Park
Champaign, IL
61820

Library of Congress Cataloging-in-Publication Data

Name: Eger, John M., author.
Title: Untold power: the marriage of art and technology / John M. Eger.
Description: Champaign, IL: Common Ground, 2022. | Includes bibliographical references. | Summary: "Most of us know that the widespread use of robotics, particularly artificial intelligence robots, will most likely have an adverse effect on the workplace and that the new jobs that emerge will require new thinking skills that the current educational system does not provide. It is also becoming clear that communities seeking to attract and nurture those most qualified for the new jobs must also renew themselves if they are to be successful. Most important, the emerging workforce must be able to engage both right and left hemispheres of the brain in-order- to solve complex problems, in increasingly creative ways. This central imperative, has resulted in the increasing demand for both artistic and creative skills along with technological and science-based skills. This treatise makes those arguments for reinvention and while it is not yet known precisely what makes people creative, many ideas about fostering creative people and institutions are discussed. The future is now. The Covid-19 pandemic has greatly accelerated our use of technology and our responses to changes we must make in education, the workplace and the workforce that have been lying dormant for too long. At the heart of the changes we must make is the vital realization that art and technology are the new benchmarks of the global economy, an economy where creativity and innovation are shaping a new world order. We are entering a new era and we must act now to prepare for a very different future"-- Provided by publisher.
Identifiers: LCCN 2021053303 (print) | LCCN 2021053304 (ebook) | ISBN 9780949313027 (hardback) | ISBN 9780949313034 (paperback) | ISBN 9780949313041 (pdf)
Subjects: LCSH: Art and technology. | Competition, International. | Creative ability--Economic aspects. | Labor supply--Effect of education on--United States. | Education and state--United States.
Classification: LCC N72.T4 E39 2022 (print) | LCC N72.T4 (ebook) | DDC 700.1/05--dc23/eng/20211208
LC record available at https://lccn.loc.gov/2021053303
LC ebook record available at https://lccn.loc.gov/2021053304

Cover Photo Credit: *Market Research Society*
Internal Images Credit: © rawpixel

Table of Contents

“The Arts can no longer be treated as a frill. Arts education is essential to stimulating the creativity and innovation that will prove critical for young Americans competing in a global economy.”

Arne Duncan, U.S. Secretary of Education, April 10. 2010

Forward

Most of us know that the widespread use of robotics, particularly artificial intelligence robots, will most likely have an adverse effect on the workplace and that the new jobs that emerge will require new thinking skills that the current educational system does not provide. It is also becoming clear that communities seeking to attract and nurture those most qualified for the new jobs must also renew themselves if they are to be successful. Most important, the emerging workforce must be able to engage both right and left hemispheres of the brain in-order- to solve complex problems, in increasingly creative ways. This central imperative, has resulted in the increasing demand for both artistic and creative skills along with technological and science-based skills. This treatise makes those arguments for reinvention and while it is not yet known precisely what makes people creative, many ideas about fostering creative people and institutions are discussed. The future is now. The Covid-19 pandemic has greatly accelerated our use of technology and our responses to changes we must make in education, the workplace and the workforce that have been lying dormant for too long. At the heart of the changes we must make is the vital realization that art and technology are the new benchmarks of the global economy, an economy where creativity and innovation are shaping a new world order. We are entering a new era and we must act now to prepare for a very different future.

Introduction

Ten years ago at a Policy meeting of the Alliance for Arts Education, I was seated a few chairs away from Melissa Shriver, Chair of the California Arts Council and wife of Bobby Shriver, R. Sargent Shriver's son and one of JFK's closest friends, talking about the role of art in education when she shocked those in attendance.

"Let's stop talking about art" she said, "let's talk about creativity because every parent, every business leader, every educator must know that's what's really important."

There is a new urgency in recognizing the role of art and the marriage of art and technology in meeting the challenges of a new creative innovation economy. While we are slowly coming to appreciate the increasing role of art as an economic and social tool as more of life and work is automated, the global pandemic has greatly accelerated these trends.

Dan Schawbel, author of *Back to Human*, wrote in a recent "AARP Bulletin." that: "Companies will more than likely invest in automation technologies as a way to limit their future exposure. And more firms will get comfortable with using video interviews technology" "after realizing the cost saving, efficacy and reach." (And) "Instead of on-site training, there will be more webinars and virtual experiences, in addition to access to online training courses." (1)

Alexandra Ossola in the" Quartz Newsletter" found that "While the coronavirus crisis has been distinctly not great for humans, it's already been a boon for robots. As humans have stayed home to slow the spread of the disease, robots have entered our lives in novel ways, from delivering groceries to taking over manufacturing jobs that put humans at risk of infection" "We have fields like telemedicine that in a matter of weeks jumped ahead to where we thought we would be in 10 years.

Robotics is being deployed into roles like policing curfews, to cleaning subways and hospitals, to delivering groceries." (2)

These investments are the trend. It isn't always so obvious but as former Google CEO Eric Schmidt, said in an interview on the CNN Special called "Rethinking Normal with Fareed Zakaria" there was a sort of silver lining for tech companies as investment in technology was likely to dramatically increase; and as many experts, critics and journalists all are saying: After the Pandemic life will never be the same. (3)

Chapter 1

On the Front Line

The use of robots in hospitals is increasing in the U.S, and perhaps among many other nations too, as Martin La Monica reported in the Conversation to "allow health care workers to remotely take temperatures and measure blood pressure and oxygen saturation from patients hooked up to a ventilator. Another robot that looks like a pair of large fluorescent lights rotate(s) vertically traveling throughout a hospital disinfecting with ultraviolet light. Meanwhile a cart-like robot brings food to people quarantined in a 16-story hotel. Outside, drones ferry test samples to laboratories.

This trend toward artificial intelligence (AI) and automation is not likely to disappear after the current pandemic is over. Baylor University of Medicine writes that "Emerging infectious diseases are infections that have recently appeared within a population or those whose incidence or geographic range is rapidly increasing or threatens to increase in the near future." (5) They note that, "The World Health Organization warned in its 2007 report that infectious diseases are emerging at a rate that has not been seen before. Since the 1970s, about 40 infectious diseases have been discovered, including SARS, MERS, chikungunka, avian flu, swine flu and, most recently, Zika." (6)

This use of robots and AI is not happening just in hospitals. Analysts predict that most current jobs in all sectors will likely be gone over the next 20 years which is why we need to better understand how the new jobs will emerge, what skills will be required and how art-based learning and art integration will help most of us better prepare for the new economy. Importantly, we need to know why art-based education can lead to creativity, and thus to innovation so that a generation of highly skilled workers emerges. This also, I hope to explain, is why robots and humans can and must coexist.

A Personal Experience

I have long been an advocate for the Arts and Arts-based Training to enhance education. The use of Robots in medicine tells us however, that as if Globalization 3.0 and the sophisticated development of Artificial Intelligence aren't threatening enough, the Covid-19 pandemic has accelerated the demand for more automation. The need for the arts and arts –based training, coupled with the tremendous role of technology, present all of us with an opportunity to meet the challenges of the new economy and the demands in the workforce.

For most of my professional life I have wondered why art and science are thought of as two separate careers, why artists and scientists as two kinds of people. Robert Root Bernstein disabused me of that when he and his wife, Michele wrote *Sparks of Genius* (7) but even before, when I was asked by former California Governor Pete Wilson to be Chair of the state's Information Technology Commission, I learned that we seemed to be having trouble in Hollywood getting workers who could draw and were also computer literate. Producers wanted an increase of H1B visas so they could look outside the U.S. for such talent.

It soon became obvious that we had a problem which was not going to be solved with more H1B visas which then gave many corporations the right to look outside the U.S.

The Demand for Talent

Root Bernstein, then a MacArthur prize fellow working at the Salk Institute in San Diego almost 30 years ago, looked at 100 or so well-known scientists over the last 200 years and found that each was equally accomplished in the arts as they were in the sciences. For example, Galileo was a poet, Einstein was a cellist who dreamed of playing at Carnegie Hall and Samuel Morse who invented the telegraph made his living as a portrait painter with several of his works hanging in the halls of Congress. (8)

As C.P. Snow pointed out in his work *Two Cultures* (9) and Root Bernstein uncovered as well, there is a symbiotic relationship between the arts and sciences that somewhere along the line we have lost. More importantly, we now know that individuals who possess the higher order thinking skills to solve complex problems prevalent today have both the muscular STEM skills as well as the creative skills that innovation requires to succeed in the new economy.

This fact becomes particularly acute when we see Artificial Intelligence systems and robots performing almost everything most people take for granted. If we are to believe the pundits. there is almost nothing a machine can't do that a human being can do. Plus they don't take coffee breaks, vacations or get sick.

What we are learning, however, is that Robots and AI systems are not very creative or innovative; those skills are still owned by humans and therein lay the opportunity and the challenge.

When I first Chaired the San Diego Mayor's "City of the Future Committee", and for 2015 to 2020 wrote for the Huffington Post about the importance of technology in cities and about creativity and innovation in education and to succeed and survive in this very economy: our schools, transportation systems, health care, infrastructure, corporate life and so forth to capitalize on those things that make us smart, those things that make us innovative, those things that nurture, retain and attract the best and brightest to a city.

Everyone says they want creative people. Getting them and nurturing them is another matter. Even today, art and art-based training is not taken seriously nor seen as a panacea to meeting the challenges of the new economy.

Even today, art and art-based training is not taken seriously nor seen as a panacea to meeting the challenges of the new economy. Rather, they are nice to have but not necessary. But as I mentioned, more and more academics and experts are starting to rethink the new thinking skills we will need in the new, creative and innovation economy

The Signs are Everywhere

Richard Riley, former Secretary of Education in the Clinton administration once observed: "The top 10 jobs in-demand in 2010 didn't exist in 2004. We are currently preparing students for jobs that don't yet exist (although some schools are trying) using technologies that haven't been invented in order to solve problems not yet seen as problems." (10) We are now at the tipping point.

Having served from 2017 to 2019 as the first Zhan Professor of Creativity and Innovation, I interviewed over 100-faculty members at San Diego State University, reviewed the entire curriculum and researched as much literature as I could find about the digital economy and the skills we needed. I recommended major changes in the curriculum and the university as well. I also visited the D School at Stanford and followed the concept of "Design Thinking" (11) closely. I believe I know where major changes can and should be made.

The educational success of Finland says that there is a call for major changes in every aspect of the educational system. There are, as well, major changes the corporation and the community must make to keep talented people engaged and productive and thus, make whole regions economic powerhouses. Other nations are reexamining their educational system as well.

Fortunately, there are successes here in the US and elsewhere to draw on. The state of California in 2019 created a Blueprint for Creative Schools (12) that looks promising. Europe's efforts to invest in more arts- based training also bears examination. Many schools have already adopted STEAM, adding art to STEM, (science, technology. engineering and math). These schools, too, should be examined closely for best practices.

The late Sir Ken Robinson, a British scholar who became an advisor to governments, educators, parents and politicians everywhere and whose TED talk (5) almost a decade ago has now been seen by millions, believed schools have killed creativity. Many of us in education have seen the enormous possibilities of reversing this trend.

Likewise, many cities where most of us spend more than half our waking life, have started or are well into strategic plans to be a smart city using technology to transform the delivery of vital public services. Some are also well aware of the need to polish off their art and culture facilities, plan sites for art and cultural districts, and afford more unique opportunities for people to meet and exchange ideas and collaborate. Good ideas come when people rub shoulders with one another.

This book seeks to tell that story, the power of the arts in the new digital economy, in a way that the body politic, educators, politicians and corporate executives understand so they begin to change the educational process and give our young the higher order thinking skills that will lead to success and survival in the new economy.

References

Blueprint for Creative Schools, A Report to State Superintendent of Public Instruction Tom Torlakson 2015, bfcsreport.pdf New Cities Foundation, http://newcities.org/

Bureister, Misti, From Boomers to Bloggers: Success Strategies Across Generations, Synergy Press, January 28, 2008.

C.P. Snow, Two Cultures, Cambridge University Press.

Emerging Infectious Diseases, Baylor College of Medicine, 2020.

Gupta, Sanjay, Rethinking Normal with Fareed Zakaria: CNN, May 11, 2020.

Ibid

Ibid

La Monica, Martin, Robots are playing many roles in the coronavirus crisis and offering lessons for future disasters, The Conversation, April 22, 2020.

Ossola, Alexandra, Coronavirus is automating the world even faster, Yahoo Sports, May 28, 2020.

Root-Bernstein, Michele & Robert, Sparks of Genius. Boston: Houghton Mifflin, 1999.

Schawbel, D, AARP Bulletin, May 2020.

Tim Brown, Tim Roberts, et al, Change by Design, Revised and Updated: How Design Thinking Transforms Organizations and Inspires Innovation, Harper Collins, 2009.

Chapter 2

Background

In the early 90's CALTRANS (California Department of Transportation) became aware that the state had no more room to build any more roads or bridges and funded a multimillion-dollar study to determine another way to do business other than by car in order to limit the number of people on the highway.

They and the researchers they hired recognized that electronic networks would play an increasingly important role in a municipality's economic competitiveness. Telecommunications could be a substitute for transportation they argued. Thus in 1997 began a program they launched called "Smart Communities" (1), that was managed by the International Center for Communications at San Diego State University. The program defined a "smart community" as "a geographical area ranging in size from a neighborhood to a multi-county region whose residents, organizations, and governing institutions are using information technology to transform their region in significant, even fundamental ways."

A slick brochure and a "How to Guide" was sent to every city manager and Mayor in the state. We were ready to assist but no one called, nothing happened. Even today, something has prevented most plans from taking shape. Hopefully, this will change as cities get more in tune with the possibilities of using technology to provide public services faster, cheaper, better.

For me, what started as a relatively simple request from CALTRANS to find a way to get more people in California to use telecommunications technology to work and shop became an eye- opener. The Smart Community program's fundamental premise was that smart communities were not, at their core, exercises in the deployment and use of technology, but in the promotion of economic development, job growth, and an enhanced quality of life. In other words, technological propagation in smart communities wasn't an end in itself, but only a means to a larger end with clear and compelling community benefit.

It's been 26 years since Tim Berners-Lee gave us the Worldwide Web (2), which made the Internet usable to the masses. In no time the great global network of computer networks blossomed from an arcane tool used by academics and government researchers into a worldwide mass medium, poised to become the leading carrier of all communications and financial transactions affecting life and work in the 21st Century. Today, with almost 5 billion million-plus users worldwide (3), growing at 10 percent per month, it is integrated into the marketing, information, and communications strategies of nearly every major corporation, educational institution, political and charitable organization, community, and government agency in the United States and around the world.

Technology Skyrockets

No previous advance not the telephone, the television, cable television, the VCR, the facsimile machine, nor the cellular telephone has penetrated public consciousness and secured such widespread public adoption this quickly. In the wake of this digital revolution, public policy and regulation is still evolving as is understanding the practical uses of the technology, the skills needed to function in the new economy, or simply, what education and training will be needed to function in this new age.

Where was this all leading? Predictions ranged from electronic "virtual communities" in which people would interact socially with like-minded Internet users around the globe, to fully networked homes in which electronic devices and other appliances whir to life on the homeowner's spoken command. From Bill Gates' book, *The Speed of the Internet* (4) and pop-scholars, like Megatrends' John Naisbitt (5), futuristic and business leaders alike painted a future that looks a lot like science fiction except that it has become a reality.

In recent years, it has become fashionable to refer to the domain in which Internet-based communications occur as "cyberspace" an abstract "communications space" that exist both everywhere and nowhere. But until flesh-and-blood human beings can be digitized into electronic pulses in the same way in which computer scientists have transformed data and images, the denizens of cyberspace will continue to live IRL ("in real life") in some sort of real, physical space a physical environment that will continue to dominate and constrain our future lives in the same way that our homes, neighborhoods, and communities do today.

Smart Communities, Smart Everything

Communities and nations around the globe often, without being consciously aware of it, are continuing to sketch out the first drafts of the "cyberplaces" of the 21st century. These are the communities looking to technology to improve the delivery of public services and in some cases, seriously looing to attract the best and brightest of worker, and corporations looking to relocate and perhaps to find those with the new thinking skills.

Singapore launched its IT2000 initiative, also known as the Intelligent Island Plan (6). Japan built an electronic future called Technopolis (7), an aggressive effort to place personal computers on every desktop and in every home in the country. And in the United States, the Clinton Administration pursued a vigorous National Information Initiative, or NII (8) one of whose early goals was to link every school and every school child to the Internet by the year 2000. Obviously, this did not happen as the Federal government envisioned.

Many communities in the United States and indeed worldwide have now undertaken similar initiatives. Stockholm, Seattle, and Sacramento, for instance, have constructed large-scale public-access networks that residents can use to obtain information about government activities, community events, and critical social services like disaster preparedness, child abuse prevention, and literacy education. The university town of Blacksburg, Virginia, has transformed itself into an electronic village (9), in which the majority of the town's businesses and residents are connected to the local data network. And cities like San Diego, as part of its "City of the Future" project, continue to build even more sophisticated electronic infrastructures that; one day soon, will allow a wide variety of local government, business, and institutional transactions.

Technology, Culture, and Place

One of the main reasons we believed over two decades ago that information networks could have a profoundly transformative effect on people, businesses, and communities was that every other major technology advance that has shrunk space and time also has remade society in fundamental and important ways.

Transportation, for example, over the millennia has done more than perhaps any other technological advance to bring the world's people closer together. But telecommunications developments, including telephones and their more modern kin, accentuated the trends inaugurated by transportation advances in three slightly different, but very important ways.

First, by allowing for rapid communication between distant sites, they made it possible for business and social relationships to flourish over long distances, permitting workers and investment capital to migrate to the most desirable locations and those with the highest economic return. Second, they extended the reach of these economic, social, and other relationships far beyond national borders, creating what was truly a global economy. And third, and perhaps most significantly, they made possible for the first time the nearly instantaneous transmission of information, collapsing both space and time in a way that no other previous technological advance had done.

The Internet, the World Wide Web, and their successors are likely to produce consequences that are as great or greater than anything we have seen so far and that are apt to be equally unexpected. If we are to maximize the positive contributions of these new technologies while minimizing their negative ones, we must begin to appreciate now how these technologies are likely to affect our people, our culture, and our perceptions of place in the years to come.

There are a few general trends worth noting.

The first is the growing ubiquity of telecommunications networks. Because it is based largely on the existing telephone system, the Internet today spans the globe, with its tentacles reaching into more than half the globe and connecting, in one form or another, an estimated five billion people. This expansion shows no signs of letting up. Indeed, as the Internet migrates from its almost purely copper-based telephone platform to cable, satellite, and digital cellular systems, the methods of connecting to the Internet will proliferate, access costs will decline, and the number of users will skyrocket.

The second general trend in the development of the Internet is the rapid expansion in bandwidth. In its original incarnation (which lasted for more than two decades), the Internet was primarily a low-volume text-based medium, and so required little transmission capacity.

The emergence of the World Wide Web, with its heavy use of graphics, photographs, and animation, changed this equation dramatically, and even top- of-the-line modem technologies at speeds from 25Mbs to 2Gps proved inadequate to the task of transporting these billions of bits of graphical information, causing many parts of the Internet to react like a two-lane freeway suddenly jammed with a hundred- or thousand-fold increase in the number of vehicles.

None of this means that the entire world's seven billion people will be hooked up to the Internet by the end of 2030. What it does mean is that the potential for connecting to the Internet is essentially unlimited and that, for an increasing share of the Internet-ready population, users will be able to send and receive not just text and simple graphics, but broadcast-quality video, audio, advanced computer graphics, and virtual reality. And they will be able to do so, not with the long waits that are common over the World Wide Web today, but nearly instantaneously with new 5G technology.

The third and perhaps most important trend in the development of the Internet is the proliferation of access points. In the past logging on to the Internet required a fairly sophisticated computer, costing in the neighborhood of $2,000 or more, an obstacle that had priced the Internet out of the range of a large share of low- and middle-income families in the United States, not to mention the vast majority of the rest of the world's population. As a public utility, as I have argued, access prices in some countries are mostly affordable and universally accessible. These nations have argued that Internet service is a right not a privilege.

Changing Geopolitical Context

These technological changes are taking place at the same time that the world's geopolitical landscape is being radically redefined. No longer dependent upon national governments for policy ideas and information, no longer content to be bound by the one- size-fits-all pronouncements of national legislators, local leaders are taking social and economic matters into their own hands, pursuing policies that will promote job creation, economic growth, and an improved quality of life within their regions regardless of the policies enacted at the national level.

This "reverse flow of sovereignty," in which local governments are assuming more responsibility than ever before for their residents' well-being, has come about at a time when information and markets of all types are becoming increasingly globalized. News, currency, and economic and political intelligence not to mention products and services making it difficult or even impossible for national governments to influence political or economic conditions over which, not long ago, they held unquestioned control. The result is a geopolitical paradox in which the nation-state, too large and distant to solve the problems of localities, has become too small to solve the borderless problems of the world.

Locally based companies that once competed with firms only in their own area code, for instance, now must battle companies throughout the world for their customers' loyalty and dollars; local governments that once had to compete for high-value residents against only nearby municipalities and the amenities they could muster now must struggle to attract such residents in a world where a growing number of people can live nearly anywhere they want and still have access to the same jobs, the same income, and the same products and services to which they have grown accustomed.

To meet these challenges, many far-sighted localities have begun to transform themselves from fractured, often highly contentious regions in which a thousand interests compete for larger shares of a shrinking economic pie into something more akin to the city-states of old than to the archetypical municipalities of modern-day political science texts.

Those that have succeeded, like Silicon Valley (also called Smart Valley) (11), possess a number of common features. One characteristic is collaboration among different functional sectors (government, business, academia, non-profit organizations, and others), and among different jurisdictions within a given geographical region. These "collaboratories" are fast becoming the new model for successful urban organization in the global age, and the only local political arrangement likely to make it possible for besieged municipalities to survive intense global competition. This point, admittedly a subtle one, is often lost in discussions of building smart communities and even in the implementation of many of the smart community projects themselves both in the U.S. and other nations too. But it couldn't be more important.

Technological Mandate

It is here that telecommunications and information technology the force behind localities' current geopolitical and economic predicament can also be their salvation. Technology has erased the barriers of time and space that physical geography has long imposed, by giving a municipality's residents and businesses round-the-clock access to information that enhances their lives, their prosperity, and their well-being.

These are just some of the many possible ingredients of the technologically driven smart communities of this century. They are the basis on which communities around the globe are competing for high-value residents, jobs, and businesses. They are also apt to be one of the most powerful ways in which financially strapped localities are able to reduce the cost and burden of government while simultaneously increasing the quality and level of government services.

However, in this new competitive environment, communities everywhere must develop a coherent and compelling vision that makes it clear how the new information networks are going to promote job growth, economic development, and improved quality of life within the community; and communicate that vision broadly. With the pandemic we also see the urgent need to reinvent learning.

The Demand for New Thinking Skills

But you cannot have smart communities, smart corporations, smart anything if you don't have smart people. Therein lies the problem. Technology alone, no matter how sophisticated its design, will not be sufficient. Smart people who have the skills to drive the new economy, to create a regional or local vibrant culture will have to have think differently, be sensitive to the new world economy and deeply know how to create and innovate.

We know that the technology everywhere and particularly broadband infrastructure could make a city smart, extremely productive and every service better, faster and cheaper than anything we could want. But few understood? It's now 2021 and we see cities around the globe scrambling to get on board. But again you can't have smart cities without smart people. Every city, every corporation, every non-profit, schools, hospitals etc. have to be smart and people need the new thinking skills to be part of these enterprises. But how do you make people smart? That is a big part of what this book is about.

Smart people in today's parlance means people who have the higher order thinking skills we talk about, people who can think outside the box, are entrepreneurial, see and feel the world and its inhabitants differently from others (certainly differently than machines), and are creative and innovative. But how do you make someone creative and innovative? There are a lot of things our educational system can do and even a few things you can do yourself.

We now know a lot more about the brain; about the importance of creativity and how it leads to innovation and the need to provide a multicultural curriculum that's real world, and hands on or as it is called, active learning. I hope to explore some of the new thinking about learning and learning how to learn, particularly about creativity and the arts, and as Stanford University developed it, "Design Thinking" as it is now called.

Artificial Intelligence, and algorithms, 5G wireless technology the Internet of Things (IoT and IoE), grid and cloud computing, fundamentally accelerated the speed and pace of all our many applications, and thus made life more complex and urgent. What we do with our high-speed riches, how we succeed and remain human were not until today bedrock issues.

In the chapters that follow, I hope to briefly talk about the so-called new economy, the challenges we face, particularly what we must do in education and in our communities to nurture, retain and attract the future knowledge workers who live there, what communities must do to best use the technology and give citizens the places they will need and be nourished by. I will also emphasize the importance of working with AI and robotics to marry art and science to give more people the "new thinking skills" they need to survive in the new economy and underscore the new man /machine interface we must learn to embrace.

REFERENCES

Brown, Justine, Smart Valley: A Network of the Future, Government Technology Magazine, September 9, 1995.

Choe, Chin Wei, IT 2000-Visions of an Intelligent Island, Elsevier, 1997.

Eger, John, Globalization A New Urgency to Building Smart Communities. Fowler School of Business, SDSU, 2005.

Gates, Bill, The Speed of Thought: Succeeding in the Digital Revolution, Warner Books, and March 1996.

History of the Web, World Wide Web Foundation. 2020.

Knesse, Tamara, What Happened to the Most Wired Town in the 1990" s, The Verge, December 20, 2010.

Number of Internet Users, Statista, 2020.

Naisbitt, John and Doris, Megatrends, Harper Collins, 1982.

Revolution in the U.S. Information Infrastructure, National Academy of Engineering, 1995.

Shapiro. Philip, Masser, Ian, Edgington, David, Planning for Cities and regions in Japan, Google books, 1991.

Chapter 3

A Creative and Innovative Economy Emerges

During the Bill Clinton Administration campaign and subsequent democratic strategist James Carville was fond of saying, "It's the economy, stupid" when talking about the singular focus of the governments' policies. (1)

Much the same could be said today. Except it's not the old economy. It's the new economy. New because we haven't yet figured out what to call it but more, and more we are coming to recognize it is a whole new economy based not on manufacturing or even service provision, but on knowledge, creativity and innovation.

Manufacturing and service provision from many nations are finding the lowest costs in other nations and seeking distribution throughout the world market. This is what author and *New York Times* columnist Thomas Friedman means when he says, *The World is Flat*. (2) Every nation, every community, every person is competing with every other. Indeed, all the economies in the world are now knit together and competing for the new knowledge jobs, the dollars they offer and the enhanced quality of life such jobs create.

We now know a lot more. Clearly our schools and the educational curriculum must change and our communities, too, where young people spend more than half their working lives and where their families, friends and fellow citizens live and work. Indeed, whole communities need to reinvent themselves.

The crucial point is that governments must come to recognize that stimulus funds and all the federal policies in the world will not help if all we do is prop up the old economy. Rather it is the new economy that is begging for attention, from nations, from cities, from corporations. Indeed, everyone and every organization is ripe for renewal or reinvention.

Creativity and Innovation

This innovative economy represents America's future. We know that educating our young people to have the new thinking skills to become productive members of the new creative and innovative workforce is vital. Yet very little is happening in that regard, in education or in the cities or at the state or national levels.

Sadly, we are just now becoming aware of how much the world has changed and how much we must change, renew, rethink and maybe reinvent our organizations, our institutions and ourselves.

For many years I taught a class on international communication, the content of which changes significantly every year reflecting rapid developments in the world. What few of us really comprehend, for example, is the difference between International Marketing, Multinational or Transnational Marketing and Global Marketing. (3) Each requires a different view of what is happening to the world market.

Globalization is a big concept. Because of the reduced cost of transportation and telecommunications, corporations continue to do more and more manufacturing around the world. Services too, are provided where the costs are lowest. Now that the world has become one market, standardization has become ubiquitous. Huge savings are achieved because the goods and services have a commonality. These is no need to tailor products to meet the differences country to country. As the marketers like to say: "one sight, one sound, one sell". The new economy is digital and Internet-based. It is changing work and altering the workplace, as we know it.

According to Nayan Chanda, director of the Yale Global Online (4), globalization really started about 50,000 years ago as people migrated from Africa in search of the unknown, to find spices, or just to proselytize others. You may wonder what 3.0 is, or 2.0 or 1.0. As Friedman puts it:

- Globalization 1.0 was from 1492 or Columbus' discovery of the new world until the 1800s.

- Globalization 2.0 was from 1800s to 2000 and it brought the world from medium to small because companies were going global. For Example the East India Trading CO.

- Globalization 3.0 from 2000 to the present -changed the world from "small to tiny." Individuals going global caused this change. The Internet was how individuals and communicate with one another.

Globalization 3.0 is here and the technology that made it possible is changing life and work as never before. As Friedman puts it, we went from a size large, to a size medium to a size tiny in just the last 20 years. This shift in connectivity has caused a radical change in employment along with devastating effects on the workplace.

As a result of globalization and advances in technology however, as Oxford University reported, "Nearly half of all jobs are vulnerable to machines to applications using information technology." (5) Oxford also predicted that more than 47 percent of the jobs that exist today in 2020- would be gone forever over the next 20 years.

How do you prepare to minimize these adverse impacts on our young people with an educational system we know is broken and with cities that are still woefully deficient in using available technology to provide vital public services? We do it by joining hands in reinventing communities that nurture, retain and attract the creative worker that author Richard Florida so eloquently outlined in his book, *The Creative Class*. (6)

The late Daniel Bell, a former Harvard professor and author of *The Coming Post- Industrial Society* (7) studied how the cotton gin helped to mechanize the farm and how the computer launched the Information age. Today the Internet and the worldwide web, robots and algorithms are dramatically and inexorably ushering in another revolution today.

The World is Digital

We don't yet have a name to describe the age in which we are living. It is digital. It is global. It is Internet based. We haven't yet begun to see the effects of grid and cloud computing, 5G, or the Internet of Everything where almost everything in the world will be tied to a world grid through sensors.

Geoff Colvin, senior editor of "Fortune Magazine" also made clear "while it seems like common sense that the skills computers can't acquire will be valuable, the lesson of history is that it's dangerous to claim that there are any skills computers cannot eventually acquire. The trail of embarrassing predictions goes way back." (8)

Yet, what Colvin and other experts seem to agree on is "that the most effective groups are those whose members most strongly possess the most essentially, deeply human abilities—empathy above all, social sensitivity, storytelling, collaborating, solving problems together, building relationships. We developed these abilities of interaction with other people, not machines, not even emotion-sensing, emotion-expressing machines." (9)

With those skills, empathy in particular, imagination, creativity, innovation and human connection, one can have transformative powers. In fact, says Ellen Caldwell, writing for JSTOR Daily, "Educators have been using the arts and humanities to teach empathy for some time. In 2003, Lauren Christine Phillips tracked the ways in which she set out as an art teacher to nurture empathy. Phillips said art students "learn how to treat others with respect, work together to solve problems, and be a part of our community. They learn this from adults at our school, as well as from each other" (10)

How did this latest phase of globalization happen so fast? Blame the Internet. At its inception the Internet was initially, simply a way for the US Department of Defense to send messages without fear of interception. (11) They used what is called "packet switching" instead of "circuit switching." Thus, by sending each message in smaller packages along undetermined routes they made interception more difficult.

With the development of the World Wide Web and the transfer of the Internet from the Defense Advanced Research Project Agency (DARPA) (12) an agency of the Department, the modern-day Internet took off. We now have close to about three billion web sites growing at about 15% per year.

By most estimates, the financial markets funded almost any company that used the word "online" or "Internet (13) because investors predicted that space was where the new wealth was and purchase of these company's stocks was a must even though they had no earnings. In 2000 investors poured over 50 billion dollars into so-called dot.com stocks and from 1996 to about 2002 about two trillion dollars. Then the bubble burst as speculators lost almost everything and were forced to sell at pennies on the dollar.

The world was getting wired, but the wires were not connected to every house, hospital or business as predicted. Other companies and investors came in and over a short time the world was wired. Globalization 3.0 was inevitable and now, in a matter of a few years, people were able to communicate with one another almost everywhere in the world.

For corporations an opportunity to reach a word market with a single message provided tremendous efficiencies.

New Laws and New Public Policy Emerges

For first time since 1934, in 1996 the U.S. legislature also passed a new Telecommunication Act (14). The Act represented a major change in American telecommunication law and was the first time that the Internet was included in broadcasting and spectrum allotment.

The Act (15) was designed to allow fewer but larger corporations to operate more media enterprises within a sector and to expand across media sectors through relaxation of cross-ownership rules. This enabled massive and historic consolidation of media in the United States and amounted to a near-total rollback of New Deal market regulation.

The Act also opened up competition by removing regulatory barriers to entry. Indeed, the conference report by the legislature describes the bill as one "to provide for a pro-competitive, de-regulatory national policy framework designed to accelerate rapidly private sector deployment of advanced information technologies and services to all Americans by opening all telecommunications markets to competition" (16).

With all these newfound regulatory riches, online companies grew overnight and social media services like Facebook and Twitter exploded. However, the problems of privacy, hate speech, spam, and fake news rose anew with little regulation in place to control these behaviors. A creative and innovation economy is emerging, but we are behind in recognizing its importance as a matter economic or public policy. We are way behind in recognizing the role of the arts or art-based instruction for the new economy.

References

Bell, Daniel, The Coming of Post-Industrial Society. New York: Harper Colophon Books, 1974.

Caldwell, Ellen C., Can Art Help People Develop Empathy? JSTOR Daily, January 16, 2018.

Chanda, Nayan, Bound Together: How Traders, Preachers, Adventurers, and Warriors Shaped Globalization, Caravan Books, 2007.

Colvin, Geoff, Humans Are Underrated: What High Achievers Know That Brilliant Machines Never Will p.40, Penguin Random House, 2015.

Featherly, Kevin, ARPANET, Britannica, May 11.2016.

Florida, Richard. The Rise of the Creative Class. New York: Basic, 2002.

Friedman, Thomas L., The World Is Flat: A Brief History of the Twenty-first Century, New York: Farrar, Straus and Giroux, 2005.

Ibid

Ibid

Ibid

Ibid

Ibid

McCabe, David, Bill Clinton's telecom law: Twenty years later, The Hill, February 7, 2016.

Read, Ash, Everything You Need to Know About Global Marketing Strategy, SUMO, January 21, 2020.

Rotman, David, How Technology is Destroying Jobs, MIT Technology Review, June n12. 2013.

Seltzer, Julian, 'It's the economy, stupid' all over again, CNN, May 8, 2020.

Chapter 4

The Role of Art in the Innovation Economy

Only recently has there been evidence of the connection between education and appreciation of the arts and success in the post- industrial information economy and society. However, the evidence is beginning to mount. Slowly, perhaps too slowly, the view that arts play a role in education is changing. In 2010 Arne Duncan, former U.S. Secretary of Education said: "The Arts can no longer be treated as a frill. Arts education is essential to stimulating the creativity and innovation that will prove critical for young Americans competing in a global economy." (1)

As mentioned earlier, the Root-Bernstein's and many others looking closely at the nexus between neuroscience and education and coming to realize that we have to find a way to marry the arts and sciences if we hope to produce more "whole brained people" with the new thinking skills the new economy needs. They realized there were two groups of people: artists and scientists and argued we should and can be both, not one or the other. In every field, inventive thinking originates in nonverbal, non-logical forms. (2)

The conclusion among scientists and others is that all students should be given early and ongoing stimulation of aural, visual and other senses and be taught to imaginatively recreate sense images. They should learn to abstract, empathize, analogize and translate intuitive forms of knowledge into numbers, words vision, sounds and movements. In some instances, feeling and sensing are communicated most naturally as literary, visual or musical expressions. Undoubtedly, the arts in a liberal arts curriculum are important in that they provide the most effective and, in some cases, the only exercise of many tools of thinking —both in expression and imagination.

The Bernstein's believe the goal of education should be understanding rather than merely knowing and the active process of learning rather than passive factual acquisition, should be its focus. For example, "it is possible to know about the principles of physics or literature without having to use them; however, being able to use them is not possible without an understanding of how they function in nature and human affairs. Students must go beyond the mere analysis of the products of creative understanding, such as poems, novels, theories, experiments, dances, paintings and songs. They must imitate and copy them to gain an understanding into the synodic and sensual processes that led to their invention." (3)

While events around the world seem to contradict the thinking that the elements are in place for the advance of the Creative Age, a period in which free, democratic nations thrive and prosper because of their tolerance for dissent, respect for individual enterprise, freedom of expression and recognition that innovation, not mass production of low-value goods and services, as the driving force for the new economy. These are the ingredients so essential to developing and attracting the bright and creative people needed to generate new patents and inventions, innovative world-class products and services, and the finance and marketing plans to support them.

Arts and sciences often interact in very productive ways that too often are overlooked, thus the glaring need for multidisciplinary education that lifts the arts onto an equal footing with the sciences. Starting in kindergarten and progressing through higher education, all students should study the arts as completely as the sciences, the humanities and mathematics. This would entail reversing the marginalization of the arts in K-12 and colleges across the country.

The arts are not merely for self-expression or entertainment; rather, they are disciplines as rigorous as mathematics or medicine. They possess their own bodies of knowledge, tools, techniques, philosophies and skills. Furthermore, because the imaginative tools used in the arts are critical to the sciences and humanities, they are owed support not only for their own sake but also for the sake of education collectively.

One hundred years of educational research has demonstrated that students are far more likely to retain and apply what they have learned if skills and information are taught as generally useful, rather than as unique solutions to unique problems. Instructors should downplay labels such as music; art and science should not only exist in insular boxes, but also rather, focus on the ways in which the same material can be used flexibly across a number of disciplines.

In order to reach the broadest range of minds, ideas from every discipline should be presented in many different forms. There is no single creative technique or imaginative skill that is adequate for all thinking requirements. Every idea can and should be transformed into numerous equivalent forms, each of which possesses a different formal expression and emphasizes a different group of thinking tools. The more ways students are able to imagine an idea, the greater their chances of insight, and the more ways in which they can express that insight, the greater the chances that others will be able to understand and appreciate it.

Harvard professor Howard Gardner, author of the seminal research and findings first published in the book *Frames of Mind* (4) discovered more than a decade ago that we learn not just through the linguistic and mathematical methods of traditional schooling but through seven intelligences: logical/mathematical, verbal linguistic, visual-spatial, bodily-kinesthetic, musical-rhythmic, intrapersonal and interpersonal. He recognized one of the primary intelligences is music intelligence and argued that music should be infused throughout the curriculum as opposed to being taught in isolation. Gardner says the same should be done with dramatic performance. (5)

Perhaps as a consequence of such research, various practical applications have been springing up throughout the country starting as early as 2010 in the poorest congressional district in the nation, New York's Saint Augustine, where the start of something profound was first reported in a PBS documentary. In a place where only one in four children once graduated, this small school now boasts that 95 percent of its students read at or above grade level, and 95 percent met New York state academic standards. All this despite a student population that was 100 percent minority, with many of the children living in single-parent homes in communities plagued by AIDS in the 1920's, crime, substance abuse and violence. (6)

What was the secret of St. Augustine's success? The school built its entire curriculum around dance, music, creative writing and visual arts. Even its science classes used the arts to illustrate research and lab experiments, while history came to life through the re-creation of events using period music and costumes.

According to Dr. Sue Snyder, (7) founder and president of IDEAS, a multifaceted organization dedicated to educational excellence in and through the arts, St. Augustine's music-oriented curriculum required every student to take music appreciation, chorus and an instrument every day, in addition to a full load of other academic subjects. The discipline and structure resulted in increasingly higher scores on mastery tests, as well as increased self- esteem.

Since that time, as Dr. Snyder has found, efforts to integrate the arts into the curriculum have taken root and succeeded in schools like Elm Elementary in Milwaukee. In the bottom ten percent in 1979, Elm is now No. 1 out of 103 schools in the district and has been for eight of the last ten years since introducing arts education. Eliot elementary in Needham, Mass., has had similar success. Since integrating art into the curriculum in 1983, test scores for average third-grade students in racially diverse schools have risen to the 99th percentile. (8)

There are many variations on the St. Augustine model of art-infused curricula in existence today across the United States and similar results are documented over and over. The arts-infused curriculum results in increased student self-esteem, attendance and enjoyment of school and often leads to higher subject mastery.

The National Endowment for the Arts (NEA) is also committed to the notion that art education needs to be emphasized in the curriculum of our schools. Starting with a $250,000 grant in the mid-1990s to the Westchester Arts council, a program called Arts Excel began to take shape. During three days of workshops at Manhattan Ville College in Purchase, N.Y., 200 teachers and school administrators learned how the arts could be used to teach history, science and math. The Council then began the process of developing a new art-infused curriculum. Over 65 individual artists as well as 21 local arts organizations helped them launch the program throughout Westchester County schools. (9)

The Massachusetts Advocates for the Arts, Sciences, and Humanities (MAASH), a statewide non-profit organization that advocates on behalf of the Massachusetts cultural community, noted in their commentary supporting the legislature's call for a "creativity index" for all the state public schools: "We have moved into an economy driven by ideas and innovation. But are we giving our students the opportunity to develop creativity-the ability to generate ideas and then to critically evaluate potential?" (10)

Maybe we really need to eliminate all the existing "silos" in education math, science, chemistry etc.- and infuse the curriculum again with the arts. As the Arts Education Partnership (11) has reported, the term "arts integration" has evolved over the past 15 years as school districts, state arts councils, and arts organizations have experimented with various models of implementation.

However, as the Partnership noted, "Some programs and schools have chosen not to use the term arts integration at all, although descriptions of the curriculum appear to belong in this domain. Much work in arts professional journals that could be termed integrative is labeled interdisciplinary, perhaps because, as noted in this review, the term evokes less controversy and challenge from within the arts professions." (12)

The Sad Truth

The sciences and math have formulas, they have equations. As Richard Deasy, former director of the Arts Education Partnership, once complained: "the fundamental problem we confront in making the arts an unquestioned part of the learning required of students and teachers is the position of the arts in the broader culture." Deasy suggested what's most valued in America is "muscularity" or toughness. The math and science curricula carry with them this sense of muscularity through their inherent formulas, truisms and theories. By comparison, the arts experience seems less tough, softer, and more anecdotal. (13)

Some business executives see the possibilities. Harvey White, co-founder of both Qualcomm Inc. and Leap Wireless International Inc., in an opinion editorial for the San Diego Union Tribune on the current state of education in 2011 wrote, "This is not an issue about including arts because it is "nice" to do so, but rather it is an imperative because our economic future is at stake". (13) Qualcomm's president since its earliest days, White was responsible for hiring thousands of engineers. Now, he says, they all need courses in art as well as science. Otherwise they will not be as creative and innovative as America needs to be in the new global economy.

White, who actually coined the phrase STEAM — for Science, Technology, Engineering, Art and Math — in a talk to the San Diego Economic Development Corporation, is especially passionate: "We simply cannot compete in the new economy unless we do something now about creativity and innovation." (14)

According to The Manufacturing Institute's National Center for the American Workforce that could change. They stated: "we believe we've only scratched the surface when it comes to the integrated arts education and STEM discussion because promising examples of integrated arts education advocacy and action are available. And assimilating STEM and art education (or art education and another business-driven education initiative) may very well be the tipping point for greater support by the business community in integrated arts education." (15)

Stem/Steam and Art-Based Learning

Clearly, federal legislation indicates we are going in the wrong direction. One example is when former president George W. Bush signed into law in 2007 a bill called the America Competes Act, (16) also known as the STEM initiative for Science Technology Engineering and Math.

The administration's bill authorized $151 million to help students earn a bachelor's degree, help math and science teachers get teaching credentials, and provide additional money to help align kindergarten through grade 12 math and science curricula to better prepare students for college. As a consequence, centers and institutes for STEM popped up across the nation. Not surprisingly, STEM is the driving educational concept. Dire futures were and are predicted unless we all get behind STEM.

In a commentary in a 2007 article in "The Wall Street Journal" Chester E. Finn Jr. and Diane Ravitch, both assistant secretaries of education in the first Bush administration, complained loudly: "This is a mistake that will ill serve our children while misconstruing the true nature of American competitiveness and the challenges we face in the 21st century." (17)

In truth, we need a huge infusion of capital and a change in attitude about art and music, math and science. We need to define a well-rounded education and to make the case for its importance in a global innovation economy. As demand for a new workforce to meet the challenges of a global knowledge economy is rapidly increasing, few things could be as important in this period of our nation's history as art and art-infused education.

The concept of art infusion or art integration is quickly bearing fruit. Former President Barack Obama called for, yet another initiative called "Race to the Top," (18) yet another STEM focus. He later better understood the importance of "STEAM," not just STEM. For example, Obama's Presidential Committee on the Arts and Humanities released a report called "Reinvesting in Arts Education: Winning America's Future Through Creative Schools." (19) It recommended, "expanding the role of teaching artists, in partnership with arts specialists and classroom teachers."

In other words, use art as a way to teach all other subjects. In fact, the National Science Teachers Association (NSTA) in a report released in 2016 said: "Teachers of science, technology, engineering, and mathematics (STEM) are discovering that by adding an "A"— the arts — to STEM, learning will pick up STEAM. Students remember science learning situations that contain multisensory, hands-on activities or experiments, which the arts can bring to science lessons." (20)

From 2018 to 2020, serving as the first Zhan Professor of Creativity and Innovation. In that role I interviewed over 100-faculty members at San Diego State University, reviewed the entire curriculum and researched relevant literature about how the digital economy was changing the skills we needed to succeed now and in the future. I recommended major changes in the curriculum and the structure of the university as well. I also visited the D School at Stanford and followed the concept of "Design Thinking" closely. I feel we know, or ought to know, where major changes in the university structure and curriculums and should be made. California State University at San Diego was in no position to do anything. Coincidentally, five college Deans, a Provost and a President resigned within 18 months and the transformational effort was shelved.

Significantly, The National Academy of Sciences, after over two years of study, in 2019 recommended that all universities fully integrate the arts and humanities into their curricula. The Academy said, "This integrative model intentionally seeks to bridge the knowledge, modes of inquiry and pedagogies from multiple disciplines the humanities, arts, sciences, engineering, technology, mathematics and medicine within the context of a single course or program of study. In such a model, professors help students to make the connections between these disciplines in an effort to enrich and improve learning." (21)

The National Science Teacher Association (NSTA) in a report released four years ago said: "Teachers of science, technology, engineering, and mathematics (STEM) are discovering that by adding an "A"— the arts — to STEM, learning will pick up STEAM. Students remember science learning situations that contain multisensory, hands-on activities or experiments, which the arts can bring to science lessons." (22). Further, Edutopia, a new non-profit think tank created by the George Lucas Educational Foundation, found that "Arts integration has been shown by several rigorous studies to increase student engagement and achievement among youth from both low and high socioeconomic backgrounds." (22)

Another organization helping reinvent the university using the arts is the "Alliance for the Arts in Research Universities" (a2ru). (23) It is one of the leading organizations to understand the power of the arts and arts integration and to see how over 50 universities are integrating the arts into higher education.

This is not art for art's sake. It is integrating the arts, or teaching all the disciplines through the arts, thus making any subject—whether it's math or science or whatever—engaging and memorable, teaching all the disciplines through the arts, recognizing the power and importance of nurturing "whole brain" people. Those in education need to emphasize how to be creative and how to solve complex real-world problems as well as get students to understand the new economy and the importance of learning how to learn, and because the world is constantly changing one can never be satisfied knowing enough. Students must acquire the new higher order thinking skills that they will not only use to graduate but to enter the new economy.

The pandemic has greatly accelerated the necessity to make major changes to how we educate, what we teach and how students learn. Higher education is at a crossroads. Colleges and universities are facing a litany of problems from severe budget deficits to fluctuating enrollment demands to spiraling educational costs to reductions in financial aid, declining student and faculty morale and to an erosion of public trust and confidence in the academic enterprise. These challenges have combined to raise the question: Is college even worth it?

Obviously, a23ru is an advocate for basic changes in learning and teaching and how the arts and arts integration are vital new ingredients to that effort. But it is also aware that change—any change—isn't easily accepted in the academy.

Faculty members who have invested years in research and are rightly proud of the sophistication and depth of their specialties are more reluctant to acknowledge the relevance of competing disciplines and, therefore, hesitant to engage in collaboration, as is usually required when doing art integration, essentially course redesign. But they too are slowly making the change in redesigning their courses.

At their conference in Denver five years ago the Alliance focused on entrepreneurship, the humanities and health and wellness noting, of course, that every discipline, every activity, every course can use the arts to talk about underlying issues or trends. a2ru recognized that Universities must make these changes if higher education is to remain the foundation of the American dream. This will take a commitment to funding and repositioning higher education and using arts integration and real-world techniques that prepare students for the new economy.

Unless the University reinvents itself for the new economy our revered institutions will suffer irreparable damage and be unable to meet the challenges of education in the next decade. Departments will continue to shrivel, be eliminated or merged; classes will continue to be dropped and staffs cut; tuition will increase, and the budget axe will continue to play havoc with its effort to be relevant.

It's not just the university either. The same can be said about high schools. An example worth examining is High Tech High in San Diego (24), another remarkable example of art infusion. HTH is a charter school well-funded by the Bill and Melinda Gates Foundation, the Gary Jacobs family (founders of Qualcomm) and many San Diego businesses. It consists of six schools three high schools, two middle schools and one elementary school with a total of 2,500 students and 200 employees. Every graduate has been admitted to college; 80 percent are admitted to four-year institutions of higher learning.

Larry Rosenstock, former CEO of High Tech High (HTH), has been accused of running "an art school in disguise." Indeed, HTH is not a school many of us would immediately recognize. It is a place with a curriculum that has turned the K-12 world upside down. Yet HTH's supporters are obviously pleased. In addition, HTH created a graduate school of education to train teachers in these new techniques.

High Tech High is unique among charter schools in that it provides personalized, project-based learning environments where all students are known well and challenged to meet high expectations. Each semester the entire faculty and student body work together on an assigned topic that draws on all the disciplines, forcing students to work collaboratively on real world problems. There are no math or art classes per se. Rather, those disciplines are infused throughout the curriculum that examines the larger questions: How does the world work? Who lives here? Why do things matter?

Rosenstock points with pride to these projects as they bring all the disciplines and all the energy and intellect of the class together, unifying the design principles of the school: personalization, adult-world connection and common intellectual mission.

Rosenstock quotes the late Sir Ken Robinson, an international expert in the field of creativity and innovation in education, who said, "Creativity is as important as literacy and should be given equal status." Maybe we really need to go back to basics to examine the true purpose of public education and what we consider an educated person to be. Maybe we need to change the vocabulary of the educational establishment, change the lenses in the camera and, in the process, awaken to the competitive demands of this new age.

Clearly, radical change is needed. Unless we dramatically alter our K-12 system of education, our young people will not find the work they want and need, the purchasing power of the average family will continue its downward spiral and the state of America's prowess in both the economic and political arena will be lost.

More educators believe as Dana Gioia, former Chairman of the National Endowment for the Arts once said: "America is not going to succeed through cheap labor or cheap materials, nor even the free flow of capital or a streamlined industrial base to compete successfully, this country needs creativity, ingenuity, and innovation." (25)

REFERENCES

America Competes Act, United States of America in Congress, Aug 9, 2007.

Art Integration, EDUTOPIA, August 29, 2912.

Bakeman, Jessica, As Race to The Top ends, controversy continues, POLITICO, July 16, 2015.

Barnes, Melody, Reinvesting in Arts Education: Winning America's Future Through Creative Schools, White House, Reinvesting in Arts Education: Winning America's Future Through Creative Schools, May 12, 2011.

Consensus Study Report, The Integration of the Humanities and Arts in Higher Education, National Academy of Science, Engineering and Medicine, 2018.

Duncan, Arne, The Well-Rounded Curriculum, Department of Education, April 10,2010.

Eger, John, Measuring Creativity in Massachusetts, Huff Post, December 3, 2010.

Eger, John, a2RU Lays Out Aggressive Agenda for Arts Integration in Higher Ed, Huff Post, November 17, 2016.

Eger, John, High school isn't what it used to be, San Diego Tribune, September 5, 2016.

Finn, Chester and Ravitch, Diane, "Not by Geeks Alone", The Wall Street Journal, August 8,2007.

Gardner, Howard, and Frames of Mind: The Theory of Multiple Intelligences Kindle Edition, Basic Books, and November 23, 1983

Gioia, Dana, Subcommittee on Healthy Families and Communities, U.S. Congress, May 8, 2008.

Ibid

Ibid

Ibid

Ibid

Ibid

Ibid

Larson, Gary O., American Canvas, National Endowment for the Arts, January 1, 1997.

National Association of Manufacturing, Skills Gap Report A Survey of the American Manufacturing Workforce, 2005.

Pilecki, Tom, et al, Something Within Me, PBS, 2001.

Root-Bernstein, Robert and Michele, Sparks of Genius: The 13 Thinking Tools of the World's Most Creative People, Houghton Mifflin Company, 1999.

Snyder, Sue, Language, Movement and Music Process Connections' in General Music Today, Spring, 1994.

Stewart, Becky, Gathering STEAM, NSTAA Newsletter, May 2014.
Strauss, Valerie, How Schools kill creativity in kids, Washington Post, April 14, 2011.
White, Harvey, "Arts and the Innovation Gap", San Diego Union Tribune, March 11, 2010.

Chapter 5

Being Creative

Einstein once said: "Everybody is a genius. But if you judge a fish by its ability to climb a tree it will live its whole life believing it is stupid." (1) The U.S. system of education, and the core curriculum, has been in existence for close to 200 years. For that matter, sadly, the whole world still uses, with some modifications of the same system over the years and the same core curriculum.

If ever there was a time to reinvent education, it is now. The world we have known it has changed dramatically with Covid-19 accelerating change and making adaptations vital to our short- and long-term success or survival. Change is hard and many ask: "Why bother?" We have managed with a broken system so far. The status quo, however, is simply not an option.

The idea and practice of universal, compulsory public education developed gradually in Europe, from the early 16th century on into the 20th. In America, in the mid-17th century. Massachusetts became the first colony to mandate schooling, the clearly stated purpose of which was to turn children into good Puritans. Other nations simply wanted kids to enter the workforce with some basic skills. So the 3R's reading writing and arithmetic- were the mandate. (2)

Education is Not Memorization

Almost anything you're curious about can be found on the Internet, and with its proliferation and the computerization of new archives and libraries on the web, literally millions of references are available with a click of a mouse. Why then is most education about memorization, when if you're interested, as Casey Stengel, of the New York Yankees was fond of saying, "you can look it up"? (3)

Most of us are well aware that we have moved to a digital world. Few of us have come to the realization that in an age of robotics, algorithms, 5G, cloud computing and the Internet of Things (IoT), we are living in vastly different world and the pace of change is unparalleled.

There is a serious debate about whether kids even need a college education to succeed, given the cost, the relevance, and the difficulty of fashioning a curriculum that provides skills for a workforce still being debated.

The prediction of Richard Riley, former president Bill Clinton's Secretary of Education that the top 10 jobs in 2020 will no longer exist in the next 20 years has even more resonance today, in large part because of the rapid-fire advances in technology that have ushered in a whole new economy and with it challenges to our communities, our workplace and our entire system of education.

At SDSU, we have known that change was necessary. Radical change. If creativity and innovation will be the hallmarks of the most successful communities in the 21St. century we need to know the answers to the fundamental questions of what makes us creative, innovative, and imaginative. But defining meaningful change is not obvious. As Zahn professor of Creativity and Innovation, I advocated "Design Thinking", a process of challenging assumptions and redefining problems in an attempt to identify alternative strategies and solutions; flipped classrooms using technology to offer 10-to-15-minute tutorials with meetings to cover points many were not getting and to answer questions (although email could usually handle many such questions), problem solving, and active learning where the students get deeply involved.

The opportunity to closely examine what we are teaching, how students learn and the challenges we faced led to my understanding of how important it is to "change the lenses in our camera" and start the process on the reinvention of education at every level.

The evidence that there is a deep connection between education and appreciation of the arts to achieve success in the new economy is mounting. The big question is whether the community, through public art, cultural offerings and a committed city council can enhance the creativity of its citizens? And if the new economy so desperately demands the creative worker and leader, what should school and universities do to prepare the next generation of creative people?

Decade of the Brain

We now know a lot more about the brain, what makes people creative and what we need to do reinvent the curriculum, reinvent the current system of education, and redefine our very definition of education to meet the needs of the new creative and innovative economy. With a joint resolution of Congress and a subsequent Presidential Proclamation declaring the 90's the "Decade of the Brain", (4) research and collaboration was widely encouraged to better understand how the brain works. The National Institutes of Health, the National Institute of Mental Health, and other Federal research agencies began comparing notes with thousands of scientists and health care professionals in universities across the country.

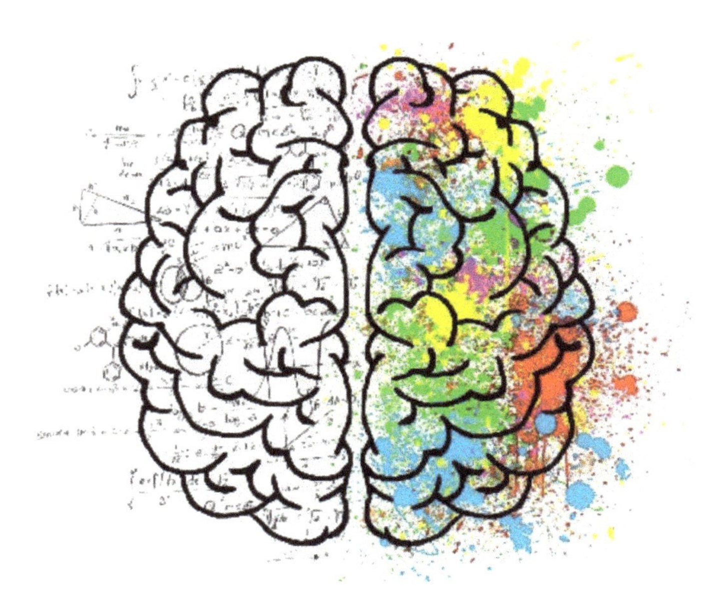

While neuroscientists do not usually characterize functions between one hemisphere and another. Artists or those trained in the arts are usually categorized as "right brained" while the left hemisphere of the brain dominates certain logical functions.

A colloquialism that acknowledges the role of the right hemisphere of the brain, according to Ian McGilchrest, neuroscientist and author of The Master and his Emissary is that "evidence shows that the right hemisphere pays wide-open attention to the world, seeing the whole, whereas the left hemisphere is adept at focusing on a detail. New experience, whatever its kind, is better apprehended by the right hemisphere, whereas the predictable is better dealt with by the left." (5) According McGilchrist and to many experts, "The left hemisphere of the human brain controls language, arguably our greatest mental attribute (while) the right hemisphere is dominant in the control of, among other things, our sense of how objects interrelate in space." (6)

Our success in a new economy demanding creativity and innovation will come from nurturing both hemispheres of the brain--the whole brain--working in tandem. Author and educator Mihaly Csíkszentmihályi calls it: FLOW a "mental state of operation in which a person in an activity is fully immersed in a feeling of energized focus, full involvement, and success in the process of the activity." (7)

Dr. Richard Restak in his book, Mozart's Brain uses the words "plastic" and "malleable" to describe the brain. He believes that we can be creative by acquiring the right series of "repertoires"; that we can "preselect the kind of brain (we) will have by choosing richly valued experiences." (8) In short, he and many other neuroscientists are beginning to conclude that we all have the capacity to be creative.

Keith Sawyer, PhD. university professor and author of *Zig Zag: The Surprising Path to Greater Creativity* that “Neuroscience and psychology have proven that all human beings, unless their brain has been seriously damaged, possess the same mental building blocks that inventive minds stack high to produce works of genius.” (9)

In his book he lists eight steps to a “greater imagination’ and details how to “build on and feed off (each step).” First, he outlines: “Ask the right question(s). And “without over thinking it, to write ten variations of the same question. For example, for the classic question: how can I build a better mousetrap, you might ask questions such as, how do I get mice out of my house? And what does a mouse want? Or How can I make my backyard more attractive to a mouse that’s my house?” (10)

The other steps he lists are:

- Become an expert
- Generate lots of ideas
- Be open and aware
- Play and pretend
- Fuse ideas
- Choose the best ideas
- Make something of your great ideas

As Sawyer further explains: “You might think of creativity only in a single context, as a quality you pull out when it’s time for a weekend craft project or a crazy practical joke. But you can use creativity to:

- Excel at your job
- Build a successful career
- Balance professional success with a deeply fulfilling personal life
- As mentioned earlier, shape your personality, your sense of style, the way you connect with the world, and the way you are perceived

- Raise your children without dull routines, harsh words, or quick-fix bribery

- Learn effectively—not by rote memorization, but in a way that makes the knowledge a part of you, so you can build on it

- Find fresh, clever, permanent solutions to nagging problems

- Make good and thoughtful decisions

- Forge interesting, sustaining friendships

- Bring about real change in your community" (11)

Art Markham, professor of psychology at the University of Texas, likes to rely on Carol Dweck, another professor of psychology from Stanford, in explaining success in school and learning to be creative. For "Fast Company Magazine" he boils it down to 3 simple rules: Be an Explainer, Be Open, Ask questions. Much like Keith Sawyer he is a strong advocate for asking a lot of questions, then rephrasing the questions into yet more questions. (12)

This looks much like what Stanford does in their D School, which they describe as "a place where people use design to develop their own creative potential." (13) They assert: "Everyone has the capacity to be creative." Many other Universities and high schools are using Design Thinking in their curriculum. IDEO, the firm that invented Design Thinking puts it this way: "Design thinking has a human-centered core. It encourages organizations to focus on the people they're creating for, which leads to better products, services, and internal processes. When you sit down to create a solution for a business need, the first question should always be what's the human need behind it?" (14)

As we enter the age of the new brain, new technologies like genetic mapping and imaging are revealing to us for the first time the mysterious secrets hidden within our skulls. The whole field of neuroscience has grown tremendously in the last few years, with continued research in the field of the arts and their role in education a recent area of emphasis. Importantly, we have learned that when both hemispheres of the brain are working together in harmony, we are more imaginative, creative and thus, productive.

Innovation Everywhere

Most analysts studying the new global economy agree that the growing "creative and innovative" economy represents America's path to a brighter economic future. Whether we can all be a Picasso or Einstein is another matter. Importantly, by focusing on a curriculum that gives young people the new thinking skills they need, we can help ensure our nation's and our children's' success in the new economy.

In 2002, a unique consortium of arts organizations embraced the principles of a study called "Authentic Connections: Interdisciplinary Work in the Arts" to enable "students to identify and apply authentic connections, promote learning by providing students with opportunities between disciplines and/or to understand, solve problems and make meaningful connections within the arts across disciplines on essential concepts that transcend individual disciplines." (15)

AS Richard Deasy suggested (16) everything is connected to everything else. The interdisciplinary curriculum "encourages students to generate new insights and to synthesize new relationships between ideas." (17) While not a manifesto for arts infusion, these recommendations go far in fostering curriculum integration and offering a way for teachers of traditional, disparate disciplines to collaborate.

The Chicago-based effort: "Renaissance in the Classroom," (18) also known as CAPE (Chicago Arts Partnership in Education), is one such model of interdisciplinary collaboration often referenced by former Secretary of Education Duncan as an example of the role art can play in schools looking for innovative approaches to learning. Such a multidisciplinary approach encourages teachers of young learners to see the connections between knowledge in one area and another, between a unit in mathematics and a unit in social studies, or between a unit in science and a unit in language arts. According to CAPE, "This process shows students that such thinking is possible and actually done in the real world." (22)

The National Science Foundation, which has the responsibility for promoting STEM, is spending a great deal of time developing grants and ideas for how to use arts-based learning to foster STEM innovation. Not surprisingly, more than 40 universities are now promoting arts integration and teaching all the disciplines through the arts, recognizing the power and importance of nurturing "whole brain" people. A National Science Foundation research firm, the Art of Science Learning, "found a strong causal relationship between arts-based learning and improved creativity skills and innovation outcomes in adolescents, and between arts-based learning and increased collaborative behavior in adults." (19)

Those in education need to emphasize how to be creative and how to solve complex real-world problems as well as get our students to understand digital innovation and the importance of learning how to learn, be entrepreneurial, be innovative and creative, solve real world problems, and understand the new digital economy. All these are part of the new thinking skills today's students need to survive.

Businesses Role

John Hagel III, co-author, along with John Seely Brown, of The Power of Pull: How Small Moves, Smartly Made, Can Set Big Things in Motion (20), made a rather telling observation that business recruiters are always looking for people who are creative, people with the new thinking skills. They noted that they look again at these people during their "exit interview."

However, finding creative people is hard work. Once they have hired these kinds of young people, they are not sure how best to use them. So it is for too many corporations.

Creativity still isn't easily measured or easy to codify, so most résumés are silent on this special qualification. In 2015 many states led by California and Massachusetts promoted the idea of a creativity index for grades K-12. There is today a "Global Creativity Index" which 25 states or nations have adopted. While admirable, it only scratches the surface. Of the measurement conundrum.

There is also a fundamental discrepancy about creativity. Business, broadly defined, seems to understand that creativity leads to innovation, but employers are not necessarily thinking of the arts as a gateway to nurturing all the new thinking skills students need. In fact, more than a decade ago, in 2007, Business Week (BW) the leading publication for business said: "The game is changing" 'It isn't just about math and science anymore (Although those are surely important disciplines) It's about creativity, imagination, and, above all, innovation." (21) Soon after publication they dropped the "Creative Age" vocabulary and started calling this new era, the "Age of Innovation." According to one BW editor who did not wish to be quoted, business readers just didn't think the magazine was talking to them anymore.

It would be naïve to think there are not barriers to rethinking any role of the arts. First, many artists are uncomfortable selling this idea of arts as a path to creativity or as a component of STEM learning. It's not what artists do, and some think it's even demeaning. Second, the arts community isn't sure how to communicate the power of arts training. Even teachers of the arts are uncomfortable championing art- infusion mainly because the visual and performing arts are electives, not required and may hurt those in higher education getting tenure. Though there are exceptions. For example, one of my university colleagues said that he "doesn't teach dance really as much as he teaches physics, and that maybe we have to find a different vocabulary to describe what we do and what the students are really learning, so the connections will be easier to make". He has a point.

It is up to arts organizations like Americans for the Arts, which is doing its part, to help change the business and philanthropic perspective. Artists and art organizations at every level need to make a stronger case that arts training leads to creativity, and that creativity can be identified and measured. Business needs to better understand the powerful role of the arts and support arts training and maybe arts integration too. Does all creativity come from the arts? Of course not, but the chances of nurturing creativity through arts-based training is a no brainer.

Last though, business executives and artists don't ordinarily mix and don't communicate with one another. A new effort to change all that might work. Just getting to know one another might be the first step for communities everywhere.

Educators need help from business and from parents too. Too often parents considering a college major for their children steer their kid away from the liberal arts and toward the hard sciences "to be sure they get a real job." Sadly, most adults are mistakenly guilty of the same when they ask kids going to college: "so what are you majoring in?" In reality, what one majors in is no indication of the kind of work they will ultimately do. While many jobs haven't been created yet they will require new thinking skills, and according to the U.S. Department of Education, some college.

Another interesting factoid is that the techno-savvy favor a non-tech education, at least for their kids and are sending them to The Waldorf School, where according to Matt Richtel of "The New York Time "the "school embraces a simple, retro look blackboards with colorful chalk, bookshelves with encyclopedias, wooden desks filled with workbooks and No. 2 pencils." (22)

What is it about Waldorf?

Developed by German educator, Rudolf Steiner, Waldorf has been around since 1919. Waldorf Education is "based on a developmental approach that addresses the needs of the growing child and maturing adolescent. Waldorf teachers strive to transform education into an art that educates the whole child the heart and the hands, as well as the head." (22)

The emphasis is on face-to-face communication and a strong personal connection between the teacher and the student. Waldorf "emphasizes creative learning with the goal of developing the child academically, emotionally and physically." They also discourage using technology in the schools and advocate (in the home too) that "television and computers (be) strongly discouraged for younger children."

The use of technology is not to be dismissed however, and while the evidence in favor of computers in schools is mixed, many experts believe that kids growing up in a digital world need the new tools of our age. They have computers and cell phones and video games, as well as Facebook and Twitter and who knows what the apps are that will demand their focus next? Thus the arguments for using technology are believed to be relevant and realistic: to reach more students, to keep more students engaged and to experiment with tools that are and will be used in today's workplace. The arguments are compelling.

At the same time many children need, and parents want, their child to have a different experience in school where they can be assured that the "whole child" and "the whole brain" is educated, where art and music are integrated into the curriculum, and where one-on-one experiences are guaranteed. At least in Silicon Valley they believe all this integrated, multidisciplinary is worth $20,000 to $25,000 in education costs. (There is ample evidence that the Waldorf experience, maybe even Waldorf, can be offered under the public system at no additional cost. But that's another story.)

The point of the Waldorf School is whether delaying the use of technology until the 3rd or 4th grade has some merit.

Dan Fost, a writer based in San Francisco has observed that "The kids don't need (technology), they learn the tech later, they build a great foundation for imagination and creativity, and computers are filling their brains with mindless junk that is often worse than anything we used to worry about from television." (23) But too much too soon also has other potential consequences according to Stephanie Brown, director of the Addictions Institute in Menlo Park, which runs an outpatient counseling and therapy program. "She's starting to see kids as young as 10 who are hooked on digital media," and Fost says, "the symptoms are strikingly similar to those of any other addiction." Brown believes this leads to "compulsivity, cravings, irritability, sleep disorders these kids build their day around their engagement with technology, and over time, they need more and more and just can't stop." (24)

Maybe using technology isn't an either/or proposition. Maybe we don't need to choose one approach over another and as we rethink the curriculum which we say we badly need to do we need to find ways of avoiding the one-size-fits-all, standardized approach.

There is a Concern

The troubling fact about creativity is that, according to *Newsweek's* lengthy article in July 2010 was that creativity grades were rapidly slipping away. Authors Bronson, Po, Merryman, Ashley for wrote that "The potential consequences are sweeping. The necessity of human ingenuity is undisputed" citing a "recent IBM poll of 1,500 CEOs identified creativity as the No. 1 "leadership competency" of the future." (25)

No one it seemed to explain why U.S. creativity scores are declining although they pointed out that the number of hours kids now spend in front of the TV and playing videogames rather than engaging in creative activities could be a significant factor. But a major concern among many experts is the lack of creativity development in our schools. In effect, they said "it's left to the luck of the draw who becomes creative: there's no concerted effort to nurture the creativity of all children." (26)

Experts are also warning, as the New York Times, found after talking with prole like the University of Georgia's Mark Runco, who says a big part of the problem is what he calls an "art bias." The Times went on to say that "Researchers say creativity should be taken out of the art room and put into homeroom. The argument that we can't teach creativity because kids already have too much to learn is a false trade-off. Creativity isn't about freedom from concrete facts. Rather, fact-finding and deep research are vital stages in the creative process. Scholars argue that current curriculum standards can still be met, if taught in a different way." (27)

Like author Daniel Pink argues, the Times went on to say "When you try to solve a problem, you begin by concentrating on obvious facts and familiar solutions, to see if the answer lies there. This is a mostly left-brain stage of attack. If the answer doesn't come, the right and left hemispheres of the brain activate together. Neural networks on the right-side scan remote memories that could be vaguely relevant. A wide range of distant information that is normally tuned out becomes available to the left hemisphere, which searches for unseen patterns, alternative meanings, and high-level abstractions." (28)

To me this is why "Design Thinking" is so critical. In 2017 The Association of American Colleges and Universities (AACU)) meeting this month (February 23-25) in Phoenix announced its upcoming General Education and Assessment Conference. The focus: "Design Thinking for Student Learning."

But what is design thinking and why now is it so important?

According to AACU, "General education sits at the intersection of an array of demands facing higher education—demands for more intentionally scaffolded, integrated, and engaged approaches to teaching and learning; more campus-community partnerships; more mentoring and advising; more multimodal learning experiences; and, above all, more meaningful assessment of student learning across these efforts." (29)

The magazine *Fast Company* says design thinking is "The methodology commonly referred to as design thinking is a proven and repeatable problem-solving protocol that any business or profession can employ to achieve extraordinary results." (30)

Forbes Magazine described it "in its simplest form, design thinking is a process—applicable to all walks of life—of creating new and innovative ideas and solving problems. It is not limited to a specific industry or area of expertise." (31)

But what is it really?

It's learning how to think and think about what you think. It's seeing a problem in its entirety, thinking "outside the box", or as academics might put it: it's divergent versus convergent thinking. Divergent thinking is a thought process that explores all the possible solutions before coming to any one conclusion. Convergent thinking doesn't explore all the possibilities but zeros in on a solution by "manipulating existing knowledge by means of standard procedures." It a quicker process, one most of us are familiar with because that's how we are conditioned to think.

Divergent thinking takes time, it usually comes after brainstorming an idea, but it has been proven to result in more creative solutions.

I asked a few of my colleagues at San Diego State University (SDSU) about design thinking. Joe Alter who teaches dance looked at paper cup while we were having coffee and asked me what w that was. I said, somewhat proudly, it's a container for holding coffee, maybe tea. He then said if you gave the cup to a 4-year-old they might build a house or a car with it.

Kotaro Nakamura an architect by profession who also heads the School of Art and Design took me through the paces when he talked about asking his classes to draw a house, and they all showed a house with a chimney, windows, a roof etc. They all looked alike. Then he asked them how many people in their family, whether they ate together, what part of the country they lived-rural or urban setting, warm weather or not-and so forth. Then he asks them to draw a house again. Obviously, it was a much different house.

As Kaihan Krippendorff in *Fast Company* called The Death of Creativity: The Death of Innovation, said: "Unplanned excursions can lead to Innovative solutions. By fostering a culture that encourages creativity, businesses can create advantage that their more rigid peers cannot copy." (32) That's why design thinking matters.

As the Internet and worldwide web enables every nation, every community and every individual to compete with every other it challenges the U.S. as never before. As outsourcing, off offshoring and the increased use of technology-call it automation-continues to change and shrink the workforce, life-long learning is the new normal, and design thinking is critical to preparing university graduates with the thinking skills to succeed in the new economy, an economy where creative and innovative problem solving is vital.

Fortunately, more and more educators are either adding to their existing curriculum, offering workshops to their teachers, or infusing every course with new teaching methods to get their students to think differently.

Art and Science are both Important

The importance of history and literature are well known if not always appreciated. The economic imperative is not as well known. Nor the urgency of rethinking the vital role of the arts, the liberal arts and humanities, as a pathway to success and survival in the new, truly global, economy and society.

But saying a college is going to create a new core where the liberal arts and humanities and all the sciences are realigned and then doing it are two different things. A criticism of traditional general education programs is that they are create too many menu options "one from column A; one from column B." The course choices may reflect the dominance of one discipline over the other or they may reflect the internal politics of an institution. They most often reflect a competition for resources as general education courses in state colleges and universities filled by hundreds of students generates a large number of dollars from the state.

In the American Association of Colleges and Universities annual survey on General Education for 2016 (33), 94 percent of the respondents reported having integrative or project-based learning requirements as an option for their students, and yet only one of four report this as a requirement for all students. If the integration is truly across disciplines and reflect STEAM or arts-integrated learning, our students will find themselves prepared for tomorrow's world. Our schools have a great deal to do, but I do not wish to place the onus on the educators. Everybody in a community needs to be concerned and involved.

The Net: Life-long Learning will be the new normal, in our schools and our communities as we continue to witness change upon change in our economy and our society. Learning how to learn and learning how to be creative and innovative is what will enable us to succeed, if not survive this bold new future. We look to educators, business leaders, politicians, policy makers and parents and community stakeholders. indeed the whole community to help us. But the question always haunting us how we can be more creative, and thus innovative?

As the Americans for the Arts (34) argued, communities can help by establishing "creative districts" in the city. Arts districts, are usually found on the periphery of a city center, are intended to create a critical mass of art galleries, dance clubs, theaters, art cinemas, music venues, and public squares for performances.

Often, such places also attract cafes, restaurants and retail shops. They can be anywhere really if the city can organize itself and others like the chamber or arts organizations and to start the process. The state can also formalize these efforts.

To date, 18 states have taken on a formalized State role in the creation of art and cultural districts. Together, they are leading the effort to transform America for the rapidly evolving creative economy.

> According to they are special areas designated or certified by state governments, that utilize cultural resources to encourage economic development and foster synergies between the arts and other businesses. State cultural districts have evolved into focal points that feature many types of businesses, foster a high quality of life for residents, attract tourism and engender civic pride."

More and more however, cities are thinking about such districts as a way to ensure the city attracts, nurtures and retains the creative workforce it needs to succeed in the new economy, an economy vitally dependent on creativity and innovation. As important as reinventing our systems of education, communities where people young and old spend more than half their day living and working, aspiring art and culture districts are essential to establishing vibrant and productive communities. Indeed, these places are the incubators of creativity.

Art and Culture Districts, says Theresa Cameron, formerly Local Arts Agency Services Program Manager of Americans for the Arts (AFTA), "have the potential with their critical mass of art galleries, cinemas, music venues, public squares for performances, restaurants, cafes and retail shops of attracting, and nurturing the creative workforce our cities need to succeed in the new economy." AFTA has created a website devoted to the "who, what and why" these districts are so important.

References

Allen, Maury, You Could Look it up, Times Books, January 1, 1979.

"Authentic Connections: Interdisciplinary Work in the Arts", Consortium of National Art Education Associations (AATE, MENC, NAEA, NDEO), 2002.

Balcaitis, Ramunas, Design Thinking models. Stanford school, EMPATHIZE@IT EMPATHIZE. DESIGN. BUILD, JUNE 15, 2019.

Bid

Brown, Tim, Roberts, Tim, et al, and Change by Design, Revised and Updated: How Design Thinking Transforms Organizations and Inspires Innovation, Harper Collins, 2009.

Bronson, Po, Merryman, Ashley, The Creativity Crisis, Newsweek, July 10. 2010.

CAPE, Chicago Arts Partnership in Education, January 2011.

Csikszentmihalyi, Mihaly, Creativity: Flow and the Psychology of Discovery and Invention, Harper Perennial, 1996.

Eger, John, America in The Creative Economy, Government Technology Magazine, August 3, 2006.

Eger, John, The Techno-Savvy Favor Non-Tech Education, at Least for Their Own, Huff Post, February 22, 2012.

Fosh, Don, Tech Gets a Time Out, San Francisco Magazine, April 1, 2010.

Goldstein, M., Decade of the Brain, US National Library of Medicine, September 1994.

Hagel, John, Brown, John Seely, and Davison, Lang, The Power of Pull: How Small Moves, Smartly Made, Can Set Big Things in Motion 1st Trade Paper edition, Basic Books, 2012.

Ibid
Ibid
Ibid
Ibid
Ibid
Ibid
Krippendorff, Kaihen, The Death of Creativity: The Death of Innovation, Fast Company, July 22. 2010.
Kuntz, Tom, American Creativity in Decline, New York Times, July 15, 2010.
Markham, Art, 3 Ways to Train Yourself to be More Creative, FAST Company, April 12. 2015.
McGilchrist, Iain, "The Battle if the Brain", The Wall Street Journal, January 2, 2010.
Pettigrew, Todd, why we should forget Einstein's tree-climbing fish. MACLEANS, July 5, 2013.
Reetz, David, Bershad, Carolyn, LeViness, Peter, Whitlock, Monica, Annual Survey of Colleges and Universities, on General Education, 2016.
Restak, Richard, M.D., The New Brain, Rodale, 2003.
Sawyer, Keith, Zig Zag: The Surprising Path to Greater Creativity, Wiley Imprint, 2003.
Smith, O.C., Little Green Apples: God Really Did Make Them! Simon and Shuster, 2003.
Staff, Design Thinking What is that? Fast Company, March 20, 2006.
Turnali, Kann, What is Design Thinking? Forbes, May 10, 2015.
www.AmericansForTheArts.org/CulturalDistricts.

Chapter 6

Art and Culture Districts: Financing, Funding, and Sustaining Them

Americans for the Arts (1) a few years ago commissioned five essays spanning the intricacies of arts, entertainment, and cultural districts specifically for policymakers, arts leaders, planning professionals, community development practitioners, and others who are interested in developing new districts or adapting existing ones.

1. Creating Capacity: Strategic Approaches to Managing Arts, Culture, and Entertainment Districts
2. Cultural Districts: Bottom-Up and Top-Down Drivers
3. Cultural Tourism: Attracting Visitors and Their Spending
4. Art and Culture Districts: Financing, Funding, and Sustaining Them
5. State Cultural Districts: Metrics, Policies, and Evaluation

These essays and reports are part of the Americans for the Arts National Cultural Districts Exchange, where you can find more information on cultural district legislation, case studies, a national district survey, and a collection of webinars. The essay here, which I was asked to write follows, with some modification on how Finance, Fund and Sustain such districts are worth reading about.

Introduction

Many cities are asking what more they can do to grow a creative economy. One way is to attract, retain, and nurture the creative and innovative workforce that is so vital in building and developing art and culture districts, particularly downtowns. For good reason, these districts and downtowns have become the "living room" for communities across the country.

An attempt to synthesize the experiences various cities have had in developing art and cultural presences is difficult because each is so different from the other. Yet, much can be learned by their experiences, and as states undertake the task of designating art and culture districts, the process of transforming our communities for the new economy accelerates exponentially.

The idea for such districts can come from almost anywhere or anyone, and it is clear that such investments cannot be made unilaterally by mayors or chambers or other leaders within the community. Rather, a successful district only evolves if a network of creative workers, art and culture and economic organizations, developers and architects, and others come together to explore their joint interests and develop a vision and a strategy for rein- venting their community for the creative age.

They may be called art and culture districts, innovation districts, entertainment districts, business improvement districts, the list goes on; but regardless of title, the goal is always to create vibrant art and culture facilities and at the same time, contribute to growing the "creative enterprises" and the new workforce that are essential to building the new economy. The reasons why such districts are popping up in cities around the world vary but, in most cases, they are designed to nurture, retain and attract the talented 21st-century workforce so vital to success and survival in the global knowledge economy.

Although everybody is talking about how innovation is what we need and will solve our jobless dilemma, few people know what innovation is or how we get it, or critically, what our communities must do to meet the challenges of the new age. It is becoming clear that art and culture districts are vital to ensuring vibrant economic activity in our cities. They are foreshadowing a whole new economy based upon creativity and innovation.

This essay will focus on the financing, funding, and sustainability of art districts and the efforts underway in the cities of San Diego, Seattle, Baltimore, Dallas, and the state of Massachusetts. Four districts or cities were chosen for a more in-depth discussion, as they seem to represent what is occurring nationwide. The state is one example of a very aggressive approach to identifying districts and helping them raise funding.

An attempt to synthesize the experiences each city has had in developing art and cultural presences is difficult because each is so different from the other. Yet, much can be learned by their experiences, and as states undertake the task of designating art and culture districts, the process of transforming our communities for the new economy accelerates exponentially.

The idea for such districts can come from almost anywhere or anyone, and it is clear that such investments cannot be made unilaterally by mayors or chambers or other leaders within the community. Rather, a successful district only evolves if a network of creative workers, art and culture and economic organizations, developers and architects, and others come together to explore their joint interests and develop a vision and a strategy for rein- venting their community for the creative age.

The sources of funding as well as continuing financing of the district, however defined, are complex and entail a mix of private and public investments in the initial funding as well as continuous financing. For example, the city or the state may provide tax waivers or incentives; a state or local designation can enable permits; economic development agencies can point to funding from yet another government agency; private investments can be made; and National Endowment for the Arts "Our Town" grants or Art Place Awards can be sought, which attracts broad-based support. There may also be hotel and tourism taxes or admission taxes that are returned to the district. All potential sources need to be explored.

A plan for continued financing to ensure sustainability of the enterprise is also important. In some cases this may mean, as in the case of Baltimore, MD, private investment each year, or tax waivers and incentives, as well as other fees generated by the district as a whole or its various parts, such as admissions from a performing arts center, art festivals, theaters, or other events within the district. It is also important that the artists, who may have lived in the art and culture district, not be forced to move as the district itself becomes the fashionable and expensive place to live and work.

Collaboration of the Whole Community

While many of the districts have already evolved over time in a some-what organic fashion, more cities are looking to art districts as vehicles for transforming the entire region. Why? Because such districts have the potential—with their critical mass of art galleries, cinemas, coffee shops, restaurants, retail shops, music venues, public art, and even office and residential housing—of attracting and nurturing the creative workforce cities need to succeed in the new economy. The new global knowledge economy depends upon a workforce with new thinking skills capable of meeting the challenge of the evolving creative and innovative workplace.

Whether such districts already exist in one form or another, or they start anew, like the IDEA initiative in San Diego, which is discussed later, the importance of multiple stakeholder involvement, collaboration, coordination, and co-location of the art and culture facilities is important. Therefore, involving all the players within a region, government, academe, business, economic and nonprofit agencies, and art and culture organizations is vital to the success of the district.

According to the Urban Land Institute, art districts, particularly ones designed to serve as incubators of creativity, are concrete evidence a new economy is taking shape. Although art districts can be essentially real estate developments—and it is important to emphasize that the private sector should be involved—they not only serve to inform, enlighten, and attract the whole community, but also represent an important economic initiative that serves the larger creative industry. The creative industry, as recognized by the American for the Arts, is one of the fastest growing sectors of the U.S. economy.

Moreover, it is becoming clear that these art and culture districts are not only the economic engines for the development of the creative industries, but for all enterprises, which them- selves must become creative and innovative to be successful in the new global economy.

While arts-based community development is nothing new. Most of the best examples do not involve the whole community. For example, Italian sculptor, painter, architect, poet, and engineer Michelangelo was a creative "placemaker" in his own right. Many artists, arts organizations, and their supporting networks have been engaged in this work for years in the United States. The foundation for the NEA's efforts began in 2007 when Jeremy Nowak, with key support from the Rockefeller Foundation and the University of Pennsylvania's Social Impact of the Arts Project, published the groundbreaking paper "Creativity and Neighborhood Development—Strategies for Community Investment," based on research in Philadelphia for The Reinvestment Fund, a Community Development Financial Institution (CDFI). This research placed arts-based work in the language and context of community investment, outlining the role of the arts in building social capital, documenting the arts as economic assets, and show how they affect market relations.

In the beginning, around the late 1980's, the National Endowment for the Arts (NEA) announced 59 new grants that will help American communities prepare themselves for the new economy. The began a plan to produce an update of the earlier report, and more importantly, a three-year effort inviting mayors and other city executives, architects, city planners, and experts in the field to "blog", and to participate in webinars and conferences to help cities and towns across America to reinvent their community for the new age, this rapidly emerging age of "creativity and innovation." (21)

Together with ArtPlace (22), a non-profit composed of the NEA and major philanthropic institutions that have traditionally funded the Arts, the nation's largest banks and eight federal agencies are helping communities reinvent their downtown and neighborhoods. In 2021 the NEA announced the first round of recommended awards for the 2021 fiscal year, totaling $27,562,040. The Grants for Arts Projects (GAP) awards ranged from $10,000 to $100,000 and cover 14 artistic disciplines, including theatre and musical theatre.

In February 2020, the NEA received 1,674 eligible GAP applications and the organization approved funding for 1,073 projects. Applications are reviewed by a panel of experts with knowledge and experience in each area under review. Recommendations are then made to the National Council on the Arts, which in turn makes recommendations to the chairman, who makes the final decision on all awards.

ArtPlace has been even more aggressive, awarding a total of $42.1 million in 134 grants to 124 organizations in 79 communities across the U.S. (and a statewide project in Connecticut). These grants, important as they are, are only part of the picture. What each state, each community city or town does to bring differing interests, differing people together is essential. For example, in just the last few years, more State Executives with responsibilities for art and culture have championed "cultural districts" to "revitalize" communities. But not clearly enough.

All these projects are, in a sense, the new incubators of creativity, by both luring the creative class to such places and by adding significantly to the establishment of a thriving creative sector, increasingly seen as a powerful economic development asset. More and more however, cities are thinking about such districts as a way to ensure the city attracts, nurtures and retains the creative workforce it needs to succeed in the new economy, an economy vitally dependent on creativity and innovation. As important as reinventing our systems of education, communities where people young and old spend more than half their day living and working, aspiring art and culture districts are essential to establishing vibrant and productive communities. Indeed, these places are the incubators of creativity.

The "Art District" designation from the Art Council, or the State seems to be enough for cities to apply, but you have to wonder what cities could do and whether smaller cities might apply if a little financial help were forthcoming. You have to wonder too, what might be possible if more organizations, chambers of commerce, economic development agencies and high-tech companies in a region joined forces to help in the reinvention effort.

A number of non-profits have also entered the scene helping cities better understand the concept of the smart city. The Smart Cities Council is a global Association with members in the U.S. Australia, India and elsewhere, offering conferences, white paper, newsletters etc. Bee Smart City, also an international association that does much the same and the New Cities Foundation, created in 2010 to help cities think through what's at stake and what's involved in the effort to renew and reinvent themselves for such a bold, new future; and "to incubate, promote and scale urban innovations through collaborative partnerships between government, business, academia and civil society." (23

Financing and Funding Initiatives

Funding the establishment of an art and culture district, including federal and state grants, private philanthropy, tax incentives or deferrals, has become extremely complex with methods that may cater to one group or another involved in the concept and planning of a district. It is important in the beginning of planning for a district that each and every organization or individual who has helped shape the district concept sits at the table to talk about where and how funding can be accessed in any way, shape, or form. Having a broad base of leaders and people who care or have control over access to resources involved in the planning process is vital since no singular agency of the government or philanthropic organization is likely to fund any district in its entirety. Rather, various pieces will have to be identified and put together to launch a district effort and continue to finance it.

According to the National Endowment for the Arts (NEA), roughly 40 percent of all funding for districts comes from private institutions; 13 percent from the government; and 44 percent from earned revenues from organizations within the district. The NEA points out that arts funding generally is a mix of public, private, and earned revenues and that continued financing depends upon donations from private enterprises and philanthropic organizations such as foundations; the NEA and Art Place (described later); and admission fees and taxes imposed on hotels and related businesses.

All these projects are, in a sense, the new incubators of creativity, by both luring the creative class to such places and by adding significantly to the establishment of a thriving creative sector, increasingly seen as a powerful economic development asset. At a time when creativity and innovation have become key indicators of success in the new knowledge-based economy, the look and feel of our cities is a vitally important measure of how a city sees itself, and how well it is meeting the challenges of a very new age. These are the essential ingredients to nurturing, retaining, and attracting the creative workforce America needs.

One of the more contentious issues is whether a specific geographic area should be zoned as an art district. Randy Engstrom of Seattle and Pete Garcia, the founder of the IDEA project in San Diego, argue against seeking formal designation. Engstrom claims that if the district area is too small, it suggests "elitism; who's in?" In other words, too, "who's out?" On the other hand, if it's too big, it becomes too diffuse and reduces demand. Likewise, Garcia avoids any formal designation because he relies, in large part, on partnering with property owners and developers who, he says, eschew regulation by the government to the extent they can.

On the other hand, according to Lisa Gegaudas of the City of Denver, even though such designations may not result in funding per se by the city, it helps marketing and funding initiatives as a matter of economic development; and it is easier to seek funding and exemptions from both the city and the state as well as from philanthropic organizations in particular, if such zoning or formal designation is given.

Placemaking Leadership

The National Endowment for the Arts (NEA) recognizes the vital link between the arts and economic development and has invested $6.5 million in more than 15 communities, to date, through its program, Our Town.

The initiative encourages private/public collaboration and creative activities to revitalize local economies. This represents "NEA's primary creative placemaking grants program and invests in projects that contribute to the livability of communities and that place the arts at their core." However, it funds other community projects where art and science disciplines are most evident and work closely with other agencies and philanthropic organizations concerned with creative placemaking.

Art Place, a collaboration between the Ford Foundation, the James Knight, Kresge, Mellon, and Rockefeller Foundation among others, has allocated more than $12 million per year for similar creative placemaking projects and represents a major step toward creative land use across America.

In a similar manner, the Rockefeller Foundation, Kresge, MacArthur, Adobe, Citibank, and other foundations provide grants supporting social innovation for improving communities and in particular, programs that enhance creativity and creative development. Kresge is very specific about its art and culture effort saying clearly that they focus "on the role arts and culture play in re-energizing the communities that have long been central to America's social and economic life. We believe that arts and culture are an integral part of life and, when embedded in cross-sector revitalization activity, can contribute to positive and enduring economic, social and cultural change in communities".

Together with the NEA's Our Town grants and the growing role of ArtPlace, it is becoming obvious that there are a number of possible organizations offering funding for art and culture districts.

Emerging Role of State Arts Agencies

According to the National Assembly of State Arts Agencies in a policy brief published in 2012, there are now 12 states playing a role in the creation of art and culture districts like Maryland, Massachusetts, and Colorado. Together, according to the policy brief, this represents 156 unique cultural districts across the country. States use a variety of tax incentives to encourage business development within local cultural districts. Examples of state incentives include sales, income, or property tax credits or exemptions for goods produced or sold within the district, or preservation tax credits for historic property renovations and rehabilitation. Maybe a state will offer an amusement or admission tax waiver for events within the district. All the plans vary.

The state of Colorado has even rebranded itself as the home of "creative industries" by changing its statewide Council on the Arts into a new Creative Industry Division of the Office of Economic Development and International Trade.

The new Division is encouraging cities across the state to apply for a state designation as an art and culture district and more recently received more than 49 applications for such designation. While the designation itself does not result in any funding, the concept, once accepted, encourages economic development, government funding, philanthropy, and so forth.

While only Maryland and Rhode Island to date have established an income tax credit, and Maryland has exempted admissions and amusement taxes on revenues earned within the district, such credits or tax deferrals are being considered by several other states within the United States. The National Assembly of State Arts Agencies continues to evaluate the emerging role of art and culture districts and has established success factors for such districts.

Local governments can also provide tax exemptions for art and culture districts as well as taxes on hotels within the downtown area. San Diego's TOT tax, which stands for transient occupancy tax, and Denver's science and cultural facilities district funding are examples of city initiatives that can fund art and culture districts or individual art and culture enterprises. Additionally, the city can include 1 percent of the estimated cost of such building projects within a city for public art and for funding other art and culture projects.

Increasingly, cities are developing the 1 percent initiative on all developments within a region to increase the funding of municipal art within the area. Developers are often requested or required by city management to provide parking, a certain number of curbs and sidewalks, etc. Public art could easily be added to this list as a way of meeting the requirement for public art or art and cultural districts.

The money paid by developers could also provide more affordable housing, which would keep artists within the district from being priced out of the area. Often the city or a government agency is required as a partner for anyone seeking funding from the National Endowment for the Arts or other similar philanthropic organizations. This encourages private/public sector cooperation.

Examples of Art and Cultural Districts in America:

Dallas, Tx: The Dallas Arts District

In the early 1970s Dallas, TX began to think seriously about relocating its art and cultural institutions into a more narrowly defined area in downtown Dallas. The city hired a series of consultants to determine how and where to house its arts and cultural institutions. In 1978, Boston consultants Carr-Lynch recommended that Dallas relocate its major arts institutions from different parts of the city to the northeast corner of downtown—an area that would allow for easy access to a vast network of freeways, local streets, hotels, restaurants, and coffee shops and become a living room for the community. With the assistance of Sasaki Associates, they "won an invited competition to develop a plan for the Dallas Arts District providing an urban design framework for private investment in and public improvements to a key 17-block section of downtown Dallas. The plan ensures that privately developed buildings and publicly financed amenities will create a lively, attractive downtown pedestrian environment." While the zoning regulations are unclear about what is or isn't within the scope of the plan, it seems to favor "compact, mixed-use development (and) accessible open space and minimized auto reliance."

With the adoption of the Sasaki Plan and the opening of the Dallas Museum of Art in 1984, the formation of the Dallas Arts District was underway. Over the next 20 years, the development of the Dallas Arts District continued with the Morton H. Meyerson Symphony Center.

In 2009, Dallas Arts District was officially created. The Dallas Arts District operates under the umbrella of Downtown Dallas, Inc., a nonprofit membership organization that serves as an advocate, steward, and representative on behalf of the Dallas Arts District. The Dallas Arts District also assumed the responsibilities of the former Arts District Alliance (created in 1984 as the Arts District Friends)—educating the larger community about the benefits and resources of the district.

The City of Dallas' public/private partnership strategy has leveraged more than $450 million in private sector investment to match the City's $149 million contribution for the development of six of the cultural facilities in the Dallas Arts District (Dallas Museum of Art, Meyerson Symphony Center, Winspear Opera House, Wyly Theater, Annette Strauss Square, City Performance Hall). Upon completion, these facilities become City property. The Dallas Independent School District also used the public/private strategy to renovate and expand its arts magnet high school which is located in the Dallas Arts District. Two cultural facilities, the

Nasher Sculpture Center and the Crow Collection of Asian Art, were privately funded and are owned independently of the City.

In addition to the cultural facilities, the Dallas Arts District has also attracted significant private investment for commercial and residential projects, most notably the 50-story Trammel Crow Center (1984), One Arts Plaza (2008), the Arts Apartments (2008), and the Museum Tower (2013).

With more than 68 acres, Dallas rightfully boasts the largest art and culture district in the United States. The Dallas Arts District is now a special zone officially recognized by the City of Dallas. All permits to build or operate within the Dallas Arts District must conform to the city's regulations. It has become the hub for new thinking about the future of the city and the entire region.

There is much discussion in the community about whether this enormous arts district will be able to create a downtown neighborhood modeled off the walkable environments so enthusiastically embraced in other cities. According to Peter Simek, editor in chief of the arts and culture site Renegade Bus, "Perhaps after decades of revisioning efforts, starts and stops, streams of consultants trying to turn Dallas into a 'real city,' the Dallas Arts District will instead give us something else, something that helps define us, something that makes us proud of this place—of what we have been able to accomplish thus far."

San Diego, Ca

The Jacobs Center

The best example of a district that grew naturally over many years is the Jacobs Center in the southern part of the San Diego. According to Victoria Hamilton, arts & humanities development manager at the Jacobs Center for Neighborhood Innovation (JCNI), the Jacobs Center was originally a gift of the Jacobs Foundation (set up by the Jacobs family). Its purpose simply was to help revitalize the neighborhood. That gift has grown into the Jacobs Center for Neighborhood Innovation (JCNI), which partnered with resident teams in southeastern San Diego to transform 60 acres into The Village at Market Creek, a LEED-certified neighborhood and vibrant cultural destination. Since its establishment in 1995, JCNI has partnered with the City of San Diego, other philanthropic organizations and corporations—and most importantly, the community itself—to create a vibrant market-driven place where the community shops, plays, and works together. Fundraising and support remain an ongoing activity.

NTC at Liberty Station

According to Alan Ziter, executive director of the NTC Foundation, formerly known as the Naval Training Center, the area consists of about 500 acres owned by the City of San Diego. When the naval base was closed in 1997, the arts community saw an opportunity in the soon- to-be-empty buildings. They worked with then-

Mayor Susan Golding and Councilmember Byron Wear to advance a Civic, Arts & Culture District in the master plan for the base con- version to civilian use.

The District is under the stewardship of the nonprofit NTC Foundation, which has raised $100 million and renovated 16 of the 26 buildings.

Rent is paid for space, with significant dis- counts for nonprofits. The City provides the buildings in an "as is" condition to the Foundation for $1 a year for 55 years. The Foundation oversees the renovation and operation of the buildings without city funding. The NTC Foundation receives $350,000–$500,000 annually from various individuals and organizations and continues to seek such gifts each year.

The property surrounding the Arts & Culture District was turned over to a Master Developer selected by the City to develop as Liberty Station. The development includes retail, office space, hotels, an education district, new homes, 125 acres of parks and open space, and historic landmarks. In addition, the Navy retained land to build 500 homes for enlisted personnel.

The NTC Arts & Culture District has several land use and zoning restrictions in place from the City and the Coastal Commission, which assures the district remains a "visitor and com- munity-serving destination." These include uses that emphasize arts and culture, first floor venues open to the public, and limits on the types of professional offices allowed on upper floors. In addition, because of the District's proximity to the San Diego International Airport, the Airport Authority has input on the building uses to comply with safety regulations. The zoning also prohibits no structures higher than 30 feet.

The IDEA District

The IDEA (Information, Design, Education, and the Arts) district in an undeveloped part of downtown San Diego called East Bay started with a plan and vision from the very beginning. Almost four years ago, Pete Garcia, artist and former CEO of an engineering firm, and David Malmuth, former development director with Disney, began promoting the importance of art and culture to economic development and focused on an under-developed area of the city that might be ideal for reinvention as an art district. With seed money raised by individuals and organizations such as the San Diego Foundation and the East Village Association, they developed a detailed strategy, vision, and plan for re-development of a 10-block area of San Diego.

With a plan and a vision in place, they spent almost two years cultivating stakeholders in the community, including art and cultural institutions, universities, design professionals, economic development agencies, and the city government. After several public forums, including charrettes, they attracted the attention of major landowners and developers in the city with whom they partnered to participate in creating the larger vision. The first phase of IDEA is an urban mixed-use project with creative office space, apartments, and retail at the street level.

Lowe's Enterprises and the West Coast architectural firm Miller Hull Partnership are IDEA's partners in developing the first phase of their plan for the

East Bay area, which will include 218 apartments intended to be "a dynamic environment for entrepreneurs and creative tech workers who want to live in the center of San Diego's growing innovation district."

A percentage, yet to be officially determined, of the apartments will include affordable housing for artists and others to live in the area. The founders of IDEA are laying the groundwork to provide up to 10 percent affordable housing to avoid gentrification.

Denver, Co

Santa Fe Denver Art and Culture District

Although there are now several art districts in Denver, and elsewhere in Colorado, the most well-known is the Art District on Santa Fe (ADSF). In large part, its popularity can be attributed to it being one of the first such districts consisting of several galleries, museums, and theaters that had organized itself as a nonprofit corporation. Since then, the district has grown to more than 60 creative industry members who pay annual dues of $300 or $360 and "friends" who pay $150.

ADSF has a first Friday "art walk" and a "preview night" for art galleries held every third Friday of the month. A small admission fee covers the costs of these well-attended events. In addition to advertising and sponsorship of activities within the district, there is philanthropic support as well. The ADSF works with other neighborhood organizations such as NEWSED and the Maintenance District to improve and beautify the area. For example, the ADSF and NEWSED split the cost of the street banners identifying the district; the Maintenance District teams with the ADSF to tackle graffiti and to beautify the streets and sidewalks.

Denver and the state of Colorado is fortunate that Governor John Hickenlooper was one of the first people to recognize the economic value of art districts when he owned a bar and restaurant in downtown Denver. Hickenlooper has said," The essence of entrepreneurship is the exploration of innovation and creativity." He understood the value that art and culture

Scientific and Cultural Facilities District

In 1989, Denver voters approved the adoption of the Scientific and Cultural Facilities District (SCFD) which, according to Lisa Gegaudas with the City of Denver Office of Art and Culture, supports art and culture, but not necessarily within the district itself. Rather the SCFD grants go to organizations and institutions and initiatives that support similar goals and ambitions in the seven-county Denver metropolitan area.

The SCFD grants advance and preserve art, music, theater, dance, geology, botany, natural history, and related activity. These funds represent 1/10 of 1 percent of the sales and use tax throughout the seven-county metropolitan area of Denver and generate substantial financial support to the districts as well as art and culture activities outside the district. According to a 2008 Colorado Business Committee for the Arts economic study, $45 million in SCFD funds catalyzed nearly $1.7 billion in total economic activity in the region.

Baltimore, Md

Ben Stone, executive director of Baltimore's Station North Arts and Entertainment District, said that the initiative for an arts district actually came from the state of Maryland, which started designating districts in cities across the state almost 10 years ago.

The Station North Arts and Entertainment District, with a diverse collection of artists live/workspaces, galleries, row homes, and businesses, spans the neighborhoods of Charles North, Greenmount West and Barclay, and is just steps away from Penn Station, the Maryland Institute College of Art, Johns Hopkins, and the University of Baltimore.

Twenty-two arts and entertainment districts have been established in Maryland since the program began in 2001. The program is administered by the Maryland State Arts Council which is within the State Department of Business and Economic Development. The Maryland State Arts Council receives applications from jurisdictions to become designated arts and entertainment districts and administers a review process to award designations. Once a district is designated, then organizations can apply for property-tax credits on qualifying ren- ovations; artists can apply for an income-tax credit when they make money on their art; and for-profit businesses can receive a waiver of the admissions and amusement tax charged by the state. Major tax benefits concerning building usage are also in place, with assessment freezes and building rehab credits to encourage growth.

The most recent economic impact study shows that in fiscal year 2013, 20 Maryland arts and entertainment districts collectively supported an estimated 5,144 new jobs that paid $149 million in wages. During the same year, new jobs combined with arts and entertainment district festivals and events supported more than $458 million in state GDP and $36 million in state and local tax revenue. The Baltimore district, which was one of the state's first designations, is intended to serve as a model for districts within the state and the nation.

The official designation helped Station North get started in earnest 10 years ago with a $50,000 grant from a local foundation. Every year since then, it has raised additional monies and operates with a budget of $400,000 annually. They have applied for additional grants from Art Place and Our Town, as well as grants from other philanthropic organizations.

The District does marketing throughout the year and also conducts surveys, evaluates media impressions, as well as comments from people living, working, and selling art and culture goods within the district. Ben Stone says these surveys help them evaluate the district's appeal.

While the district has a geographic designation, it is not zoning per se. Anyone can launch a business, a coffee shop, a restaurant, an art or culture organization in the district without any approvals whatsoever. However, if an artist would like to receive an exemption from state income tax, they must apply and be approved by the Maryland State Comptroller's Office.

A big incentive for artists who live and/or work in the area is the ability to sell their goods with an exemption from the state income tax. In addition to the income tax benefit, there are property tax incentives for the renovation of spaces used for purposes within the district as well as an exemption for revenues generated from admission or amusement for those art and entertainment activities occurring within the district.

The Commonwealth of Massachusetts

In order to nurture "Creative Communities," Massachusetts has a unique program for funding art and culture. Massachusetts' Cultural Council, which administers all art and culture programs, has made it clear it supports "projects that revitalize communities, create jobs, grow creative industries, and increase engagement in cultural activities by Massachusetts residents and visitors." In carrying out its convictions, the Council established the Adams Arts Program in 2005 to support the development of cities, towns, and neighborhoods by promoting art and culture.

They define a cultural district as "a specific geographical area in a city or town that has a concentration of cultural facilities, activities, and assets. It is a walkable, compact area that is easily identifiable to visitors and residents and serves as a center of cultural, artistic, and economic activity." To date, the Council has "invested $9.4 million in more than 100 projects statewide, from Pittsfield to Provincetown, involving more than 950 nonprofit organizations, businesses, and local governments. Adams-funded projects have raised more than $27 million in matching funds, making for a combined investment in the Massachusetts creative economy of nearly $38 million."

Applicants can apply for funding up to $50,000 but must agree to match the size of the grant with both cash and in-kind contributions of no more than 20 percent. After that, they may apply a second time, but in-kind contributions are not allowed. This usually compels the requesting organization to seek philanthropic participation.

The Massachusetts Cultural Council also requires that the applicant represents at least three organizations each with specific responsibilities and demonstrate a "systemic approach to cultural economic development," with the project "tied to clearly defined economic goals and objectives" and that the "funding plan that demonstrates project viability, community support, and the potential for long-term sustainability."

The Council has separate programs to fund facilities and support individual artists. The current allocation for facilities supports "capital projects in the arts, humanities, and sciences that expand access and education; create jobs in construction and cultural tourism; and improve the quality of life in cities and towns across the commonwealth." The programs in support of artists include fellowships, apprenticeships, individual community grants, grants to schools to support artist residencies, and even helping find affordable artist spaces.

The state keeps tabs on employment, spending on art and culture, and the contribution such spending, including the state grant programs, have on the economic health of the state. For example, a report in 2011 stated, "More than 8,000 arts and culture nonprofits in Massachusetts, which employ 27,100 people, spend $2.1 billion annually and generate another $2.5 billion of economic activity across the state."

Moreover, the report noted that the creative sector has a major impact on the larger economy and is "a steady, reliable industry, less subject to the cyclical ups and downs of the overall economy than the average New England business."

According to Meri Jenkins, program manager at the Massachusetts Cultural Council, the state knows that its "future prosperity is closely tied to the creative economy." Last year, for example, the art and culture sector accounted for "27,000 full-time, part-time, and independent contractor jobs and pumped $1.2 billion into the state's economy through direct spending."

Measuring Success

As indicated, Baltimore's district relies on occasional surveys media impact tries to measure success of individual festivals and events. Dallas does not do an arts economic impact study on a regular basis but is always mindful of the importance of metrics. Many art districts conduct evaluations as part of a larger economic evaluation or as part of a downtown survey.

While there are not many good evaluation tools available for art districts, one might want to look at Maryland's tool as one of the better ones: Maryland Arts and Entertainment Districts Impact Analysis FY 2013. This study is conducted on a regular basis by the Regional Economic Studies Institute (RESI) of Towson University. The study evaluates the economic and fiscal impact of Maryland's 20 arts and entertainment districts. To quantify the economic and fiscal impact of the arts and entertainment districts, RESI used the "MPLAN" input/output model.

ArtPlace America now uses Vibrancy Indicators to evaluate the effectiveness of its grants. According to ArtPlace, the "Vibrancy Indicators are designed to measure the non-arts place- making impacts of ArtPlace America investments based on numbers of arts-related business establishments and the presence of nonprofit arts organizations, as part of our effort to identify the most active arts neighborhood in each of the nation's large metropolitan areas."

The Texas cultural district program, like Colorado, Maryland, Massachusetts, and other states, is providing certification for districts—going one step further to identify essential factors that will lead to success promoting art districts and raising and nurturing support. Specifically, they have identified the following measures of success:

- Antique and authentic identity
- Strategic partnerships
- Artist live/workspaces
- Anchor institutions
- Events
- Clear signage
- Marketing and promotion • Strong amenities

All art districts serve, in a sense, as the new incubators of creativity, by both luring the creative class to such places and by adding significantly to the establishment of a thriving creative sector that is increasingly seen as a powerful economic development asset.

There are obviously many ways in which art districts, however they are defined, can be funded and financed. At the root of a successful foundation is private/public sector collaboration—between art and culture organizations and agencies and organizations responsible for economic development such as the Chamber of Commerce. When multiple teams join together to create a vision, a plan, and a strategy, seeking funding is shared work and will ultimately lead to success. After the foundation is built, an ongoing revenue stream will strengthen communities and inevitably attract and nurture the next generation of workers for the new creative economy.

A community venturing to create such a district must keep upmost in mind, "Art Districts", particularly ones designed to serve as incubators of creativity, are concrete evidence a new economy is taking shape. Although art districts may seem essentially real estate developments—and it is important to emphasize that the private sector should be involved—they not only serve to inform, enlighten, and attract the whole community, but also represent an important economic initiative that serves the larger creative industry. The creative industry, as recognized by the American for the Arts, is one of the fastest growing sectors of the U.S. economy. Moreover, it is becoming clear that these art and culture districts are not only the economic engines for the development of the creative industries, but for all enterprises, which themselves must become creative and innovative to be successful in the new global economy.

REFERENCES

ArtPlace http://www.artplaceamerica.org
Baltimore Station Arts and Entertainment District http://www.stationnorth.org
Colorado Creative Industries http://www.coloradocreativeindustries.org
Dallas Arts District http://www.thedallasartsdistrict.org
Denver Art District at Santa Fe http://www.artdistrictonsantafe.com
IDEA http://www.ideadistrictsd.com
Jacobs Center for Neighborhood Innovation http://www.jacobscenter.org
Maryland State Arts Council www.msac.org
Massachusetts Cultural Council http://www.massculturalcouncil.org/ about/staff.asp
National Assembly of State Arts Agencies http://www.nasaa-arts.org
National Endowment for the Arts http://arts.gov
NTC Liberty Station http://www.ntclibertystation.com
Seattle Office of Arts and Culture http://www.seattle.gov/Arts/
www.AmericansForTheArts.org/CulturalDistricts.

Chapter 7

Smart Cities

The cities and the communities where we live and work and play are essential to ensuring our individual and collective growth and renewal.

As a new economy based on knowledge use and production is unfolding, media and public policy experts are heralding the evolution of smart cities deploying technology to provide vital public services and the infrastructure demanded by the new economy. Understanding these developments requires a deeper understanding of the role of technology, particularly the Internet and the World Wide Web, the economics of doing business in the digital age, and the politics of global problem solving.

A new term has entered the lexicon: The Internet of Things or The Internet of Everything (IoT or IoE).

It has been established that almost anything can be automated. For example, with a sensor attached to blades-of-grass we can connect them to the Internet and create a lawnmower system turning a water sprinkler on or off as needed. By roughly the same method we can save energy and water, rebuild our systems of transportation, erect smart buildings and even smarter systems of education. There is no limit to what the new technology allows us to do,

Right now every city is laying fiber, the information highways of the future. As mentioned earlier, San Diego State University, working with the California Department of Transportation developed the concept of "smart communities" communities using new wired and wireless information infrastructures to connect every home, office, school, and hospital, organization and institution, large and small, to one another and through the Worldwide Web, to millions of other individuals and institutions around the world. (1)

The new broadband infrastructures coupled with Big Data Analysis provide opportunities to reinvent the city. Big Data is a term that applies to the growing availability of large datasets in information technology that once captured, stored and evaluated can serve to make city services while engaging the general public to help create. Big Data is now in abundance as more data or information is posted online. CIOs and county executives noted that solving these challenges would require partnerships with the private sector to marshal resources, talent and technological know-how. The city has neither the money nor the expertise to do all they must do to better serve their communities.

for-profit organizations using social media and other internet resources. And from the Internet sites themselves

Former Mayor of Indianapolis and Deputy Mayor of New York, Stephen Goldsmith and Susan Crawford, co-director of the Berkman Center for Internet and Society at Harvard, are strongly in favor of what they call "smart data" providing "efficiencies" that save taxpayer money" (and) "build trust in the public sector." (2) But it is clear the city knows, as it surely does, a) that it alone cannot afford to become a smart city without relying on the private sector; b) engaging the whole community; and c) that art and culture are important ingredients in the transformation to a smart city, a smart community.

Many city leaders worry about what is called "PPPs" of private public partnerships. According to a report by Thompson Reuters, local governments face "complex issues": Dealing with the opioid epidemic, reducing homelessness, providing affordable housing, and mitigating the impact of natural disasters such as wildfires and floods. All were listed as top challenges in recent research from the Governing Institute. After interviewing elected officials and chief information officers (CIOs) across six U.S. counties in June 2018, the Governing Institute found government leaders have a hard time tackling these challenges because of restrictions in how they finance projects and programs. A lag in technology adoption and difficulties attracting and retaining IT talent were also barriers." (3)

Private companies however have the expertise and the funding and are eager to partner. According to Cisco Systems, IoE could generate $4.6 trillion in value for the global public sector by 2022 through cost savings, productivity gains, new revenues and improved citizen experiences. Cities have the potential to claim almost two-thirds of the non-defense (civilian) IoE public sector value. Cities, they believe, will capture much of this value by implementing "killer apps" in which "$100 billion can be saved in smart buildings alone by reducing energy consumption. (4)

Private companies however have the expertise and the funding and are eager to partner. According to Cisco Systems, IoE could generate $4.6 trillion in value for the global public sector by 2022 through cost savings, productivity gains, new revenues and improved citizen experiences. Cities have the potential to claim almost two-thirds of the non-defense (civilian) IoE public sector value. Cities, they believe, will capture much of this value by implementing "killer apps" in which "$100 billion can be saved in smart buildings alone by reducing energy consumption. (4)

Goldman Sachs calls the IoE the 3rd wave, and points out that while, "The 1990's fixed Internet wave connected 1 billion users the 2000's mobile wave connected another 2 billion. The IoE has the potential to connect 10X as many (billion) "things" to the Internet by 2020, ranging from bracelets to cars. Gartner research says that this past year alone we have 4.9 Billion Connected "Things." (5)

PPPs seem easy but both public and private sectors have agreed to be flexible as they don't always work as planned. Miranda Spivak of "The New York Times" found that "private investors need to do more than just look at these deals as a chance to make a quick profit, said Stephen K. Benjamin, the mayor of Columbia, S.C." "This not only requires municipal and local officials to have an open mind, but it is also really going to require institutional investors to rethink the way they bring these (joint partnership) ideas" "This is about community problem solving as opposed to, 'We see this collective financial opportunity." (7)

Public Trust

Getting the community on board is perhaps one of the biggest obstacles. *US News and World Report* this said year that "when Edelman released its annual Trust Barometer, (2121) it found that Americans' trust in their government had plummeted 14 points in a single year, to 33% percent—the steepest drop the global communications firm has ever recorded. Meanwhile, a decades-long campaign backed by deep-pocketed special interest groups has convinced many that government not only can't solve their problems, but it is also the problem". (6)

Without buy in from the public and meaning involvement, doing any major projects becomes controversial. They went on to say "we've seen how the public's trust—or lack of it—affects those in office and their willingness to take risks or try something new. In low-trust communities, we hear too often that government should "stick to the basics," leaving innovation to the private sector or creative non-profits." (7)

The loss of public confidence matters. Research shows that trust in government is necessary to spur economic growth and business investment. It influences individual behavior in public campaigns, such as those promoting vaccination or getting people out of cars and onto mass transit. The authors went on to note that "In our own work with elected officials around the country, we've seen how the public's trust—or lack of it—affects those in office and their willingness to take risks or try something new. In low-trust communities, we hear too often that government should "stick to the basics," leaving innovation to the private sector or creative non-profits. (8)

As we weave more AI into IoT and IoE devices we should also increase transparency so even if human, consumers or citizens, have nothing to contribute they should not be unaware of the pros and cons of such efforts. Take streetlights for example.

In San Diego recently, proudly announced that "Approximately 4,200 intelligent sensors (were) being installed to transform the City's streetlights into smart infrastructure that will help optimize traffic and parking, plus enhance public safety, environmental awareness and overall livability for San Diego residents." (9)

As the city explained "These intelligent sensors can see, hear and feel the heartbeat of a city. The node connects city officials and citizens to real-time data, allowing for endless applications. From easier parking and decreased traffic congestion, enhanced public safety and environmental monitoring, enhanced bicycle route planning, to enhanced urban and real estate development planning, this platform can improve the quality of life in our city and boost economic growth." (10)

The only problem is the citizens of San Diego, were uninformed, they had no idea that the city made such a decision and saw it as tantamount to spying. "A web of surveillance technology" "wrote the American Civil Liberties Union, the Electronic Frontier Foundation, civil rights advocates, Muslim groups, and other groups awakening to various cities efforts to use technology to gather data." (11)

They complained, "Devices capable of monitoring and recording residents invade privacy, chill free speech, and disparately impact communities of color. Certain technologies, adopted for benevolent purposes in the name of 'smart cities,' may gather and store information about how residents live their lives in public places, including churchgoing habits, participation in political protests, or visits to an abortion clinic". (12)

The citizens made it loud and clear. They wanted construction of its planned high-tech intelligence network stopped until a "privacy advisory task force" (13) was organized. San Diego's streetlight story, it was alleged, had been plotted behind closed doors, and bereft of transparency, leading critics to question city hall's ethos of secrecy and willingness to bend ethics in the pursuit of campaign money.

Their concern is slowly gaining in importance. As early as 2014, Steven Poole of the Guardian said, "The smart city is, to many urban thinkers, just a buzz phrase that has outlived its usefulness: 'the wrong idea pitched in the wrong way to the wrong people'. So why did that happen and what's coming in its place?" (14)

Just this year "Government Technology Magazine" asks, "Is Smart City a Euphemism for plain old Surveillance" Quoting Julia Thayne, a founder of the Urban Movement Labs, "Too often, smart city projects end up being "just plain old surveillance". (15) And Computer Weekly makes it clear that "Digital security is another threat city face when they try to implement smart city projects. As personal data gets uploaded into the cloud, it is often shared with digital devices, which, in turn, share the information among multiple users. It is therefore vital to safeguard this information from unwanted use. Applying appropriate digital security measures safeguards the private and proprietary information of citizens, governments, research partners, universities and digital infrastructure". (16)

Other cities are both working to forge ahead in partnership with private companies and importantly, involving community organizations and individuals who want to help reinvent their city. For example, two years ago the city of Philadelphia released a citywide information technology strategy (17) in an effort to increase transparency and boost the city's relationship with the academic, nonprofit and private partners. It is also a guide to allow the city to lay its IT strategy bare for its residents. And, coincidentally, Philadelphia, infused art into the classrooms, created performance space galore and take their art and culture responsibilities perhaps more seriously than most cities has well thought-out plan underway

Andrew Buss, the city's deputy CIO for innovation management, said the report's contents were "informed by a series of resident-filled workshops launched January 2018. Each group targeted one of the report's categories. (18)

For the digital access and equity category, for instance, the workshops led to an acknowledgement of the city's grant-based digital literacy work, but the report also announces Philadelphia will search for a digital literacy coordinator to centralize and shape its digital-literacy policy.

To improve digital access, the report shares a plan to develop unique property and building identification numbers across city agencies, as well as a new civic-data initiative with Johns Hopkins plan, reflects the fact that IT has moved from a largely internal, behind-the-scenes department to one that has a public role and coordinates across many city departments. This is what every city, every institution should be doing to take advantage of the technology, citizen know how and expertise and support. Not just the city, but to really take advantage of the opportunities these advances represent, private /public cooperation and citizen involvement is vital.

Cities are more important than ever as nations enter headlong into the Flat World, as Thomas Friedman has written. It is a new age of creativity and innovation; an age when technology has shrunk the world and where every nation, every community, every individual is suddenly competing with every other. It is a new age where every nation, every community, every individual is suddenly competing with every other.

Unfortunately, the importance of art and culture, the role of libraries, and museums, open space performing art centers, opera halls and music venues get second billing. Yet, having welcoming and engaging city is also having a positive effect on nurturing and attracting the best and brightest.

Richard Florida, citing a recent US Bureau of Economic Analysis published in the Bloomberg CityLab Newsletter, found that "Arts and culture are key components of quality of life, and important contributors to urban economies as well. My own research finds that arts and cultural employment is one of three key drivers of urban economies—alongside science/technology and business/management occupations. Two recent studies provide a fresh look and new data on how the contribution of the arts to state and local economies". (19)

This is the reason that so many states in the US has an Art Cultural program. According to the National Assembly of State Art Agencies, such "districts are special areas designated or certified by state governments, that utilize cultural resources to encourage economic development and foster synergies between the arts and other businesses. State cultural districts have evolved into focal points that feature many types of businesses, foster a high quality of life for residents, attract tourism and engender civic pride." (20)

Arts districts, usually found on the periphery of a city center, are intended to create a critical mass of art galleries, dance clubs, theaters, art cinemas, music venues, and public squares for performances. Often, such places also attract cafes, restaurants and retail shops. How to establish and fund an arts district is vitally important.

Beginning in 2017, a new arm of the Foundation called the Global Culture Districts Network (GCDN), is focusing on the role of "culture districts" areas including museums, theaters, performing arts centers, and other assets such as live/workspaces, coffee shops and restaurants that help define a city. Think of London, New York, Berlin, or Paris.

The Global Culture Districts Network aims to ensure that these projects are vital assets for their communities, contributing to the vitality of 21st century cities, according to AEA Consulting Director Adrian Ellis, who is also a Director of GCDN, "The idea of GCDN is to support the leaders of Cultural Districts both planned and existing wherever they are," he says. "There are clearly many differences between Seoul, São Paulo, Vancouver and Muscat to take a random four but there are many similarities too, in a world where ideas, people and capital are highly mobile." (24)

In many ways smart cities alike Saadiyat Island in Abu Dhabi, Beijing's Olympic Green, Dallas Arts District, Chicago's Millennium Park, Hong Kong's

West Kowloon Cultural District, Singapore's Esplanade, Doha's Cultural District, are cited as models of how high-profile urban developments have been planned to embrace cultural activities as an important part of the public realm. In addition to an array of programs discussing the future of cities, leaders from Hong Kong, Qatar, the UK, Amsterdam, Germany and the U.S are shaping the Global Cultural Districts Network (GCDN).

When meeting in Dallas, Texas, several years ago, The New Cities Foundation attracted over a thousand leaders from cities around the world, most of whom heard about creative place making, technology innovations, new transportation schemes, sustainability, financing and audience building as well as a host of other topics all of which were discussed in a plenary session and numerous breakout groups.

Ellis argued that: "For planners (the art and culture districts) can help build community and social capital; for sociologists they keep at bay the forces of anomie; for economists, they incubate and inculcate creativity, and draw those fickle high-net-worth tourists; and for the politicians and the like, they signify and calibrate complex aspirations and identities. But they are difficult to get right, and expensive and politically embarrassing to get wrong." (25)

The GCDN is hoping to provide the following services for its members:

- Regular convening to share emerging best practices, hear expert panels, and discuss the place of cultural districts in urban policy, economic development and related areas of public policy such as travel and tourism.

- Original research on topics of common interest such as programming, audience development, cultural tourism, professional development, relevant trends in technology and creative industries strategies.

- Regular summaries and circulation of secondary research and news of common interest.

- Virtual forums for detailed sharing of information and discussion of opportunities and challenges.

- Opportunities for establishing strategic partnerships for content, programming, skills training, and knowledge transfer.

"This isn't just about the arts," Ellis said, "This is about urban policy." Whatever the motivation, cities are beginning to see the need to renew and reinvent themselves for a very complex and different world economy. "It's scary" he said, and executives, architects, city planners, and policymakers alike, must begin to collaborate on the effort of "Reimagining the City."

Public Art

In a green paper for Americans for the Arts, a leading advocacy non-profit in the US, they argue persuasively that public art is not only importance to for of a city important, essential to its economy: "As has been witnessed throughout history, public art can be an essential element when a municipality wishes to progress economically and to be viable to its current and prospective citizens. Data strongly indicates that cities with an active and dynamic cultural scene are more attractive to individuals and business." (26)

Public art can be a key factor in establishing a unique and culturally active place. Public art can create civic icons, but it also can transform our playgrounds, train stations, traffic circles, hospitals, water treatment facilities, and airports into more vibrant expressions of human imagination. By building and reinforcing community culture, public art can act as a catalyst for community generation or regeneration. In this case, size does not necessarily matter. Public art can be very visible, large, permanent and unmistakable as an art experience; but it can also be very subtle, short lived or seamlessly integrated into one's experience of a place. Public art matters."

According to Penny Bach, an arts consultant to the Fairmont Part Arts Association in the city of Philadelphia, "public art occupies a unique position within the art world. In comparison with big-name gallery shows, public art is often "under -appreciated" much like landscape architecture is. But there's lots to applaud: "It's free. There are no tickets. People don't have to dress up. You can view it alone or in groups. It's open to everyone." (27)

She also points out that a "Community art can also create attachment to one's community studies have looked at the economic development benefits of art, but only just recently have there been wider examinations of the effect of art on a community's sense of place. The Knight Foundation's Soul of the Community initiative surveyed some 43,000 people in 43 cities and found that "social offerings, openness and welcome-ness," (28) and, importantly, the "aesthetics of a place its art, parks, and green spaces," ranked higher than education, safety, and the local economy as a "driver of attachment." Indeed, the same story may be playing out locally in Philly: a survey of local residents found that viewing public art was the 2nd most popular activity in the city, ranking above hiking and biking."

Historically, Alastair Snook, writing for the BBC, say that "in a broad sense, public art is as old as the hills: think of the statues of the pharaohs of ancient Egypt. The four colossal-seated sculptures of Ramesses II hewn out of the sandstone facade of his rock temple at Abu Simbel in southern Egypt were designed with a very specific public in mind his Nubian enemies. A blunt display of imperial chest thumping, this is art that bludgeons the viewer into submission. Millennia later, Michelangelo's marble statue of David offered another example of the symbiotic relationship between art and the state: positioned outside in the Piazza della

Signoria, it became a public symbol of the independence of the Florentine Republic." (28)

In the 20th century, though, public art really came into its own, the National Endowment for the Arts, The Knight Foundation, Americans for the Arts, various businesses and others began promoting and funding what they called "Creative Placemaking", which The National Endowment for the Arts, describes as "projects (which) help to transform communities into lively, beautiful, and resilient places with the arts at their core. Creative placemaking is when artists, arts organizations, and community development practitioners deliberately integrate arts and culture into community revitalization work placing arts at the table with land-use, transportation, economic development, education, housing, infrastructure, and public safety strategies. Creative placemaking supports local efforts to enhance quality of life and opportunity for existing residents, increase creative activity, and create a distinct sense of place." (29)

Grantmakers in The Arts says public funding for the arts -Federal, state, and local public funding totaled $1.37 billion in FY2018. This is best understood they say "by tracking congressional allocations to the National Endowment for the Arts (NEA), legislative appropriations to state arts agencies, and local government funds going to local arts agencies. These entities distribute public grants and services to artists, creatives, and cultural organizations across the nation." (30) What corporations, philanthropic organs and individual contribute probably double the total figure.

Eight years ago the Obama Administration announced "a new "Smart Cities Initiative that will invest over $160 million in federal research and leverage more than 25 new technology collaborations to help local communities tackle key challenges such as reducing traffic congestion, fighting crime, fostering economic growth, managing the effects of a changing climate, and improving the delivery of city services." (31)

Not surprisingly, a whole new economy based not on manufacturing or even service provision, but on knowledge or more precisely creativity and innovation is slowly taking shape. This is perhaps what Wellington Webb, former Mayor of Denver and past president of the U.S. Conference of Mayors meant when he said: "The 19th century was a century of empires, the 20th century, a century of nation states. The 21st century will be a century of cities." (32)

Rise of the Region States

While the city reinvents itself it's important too, to be asking "What is our City" Cities have grown and morphed in ways we may not have imagined.

Kenichi Ohmae, writing in Foreign Affairs, first seemed to see this dramatic shift in power almost 20 years ago when he pointed out that: "The nation state has become an unnatural, even dysfunctional, unit for organizing human activity and managing economic endeavor in a borderless world. It represents no genuine, shared community of economic interests; it defines no meaningful flows of economic activity. In fact, it overlooks the true linkages and synergies that exist among often disparate populations by combining important measures of human activity at the wrong level of analysis." (33)

Although, the Brooking Institution, which has argued, we need to think not of the city or the state but Metro Regions, this has not yet been easy to do. About a year ago, City Council President Ben Hueso was thinking of consolidating the city and county governments of San Diego similar to the city-county model used in San Francisco.

He appeared before editorial boards and citizens groups and on every appropriate occasion made his intentions clear. Hueso's reasons were simple. He said then that such a merger of the city and the county "could reduce costs and improve government services in a tough economic climate. Since making this suggestion, Hueso received hostile reactions from the county and a lack of any noticeable public support. (34)

According to Michelle Ganon, Hueso's director of communications, Hueso could not see pursuing this idea. It was not going to be productive. Unfortunately, this is not the first time that the idea of consolidating governments has been suggested, and then dropped. (35)

According to Bruce Katz, a founding director of the Metropolitan Economy Initiative at the Brookings Institution in Washington, D.C., "The mismatch between governments and the economy undermines the competitiveness of places by raising the costs of doing business, exacerbating strong development trends, squandering urban assets and deepening racial and class separation." (36)

People live in one municipality, work in another, and go to worship or to a doctor's office in another or to the movies in a third. And yet different, fragmented governments represent all these different places.

As David Kocieniewski pointed out in The New York Times, "The crazy quilt of municipal governments that ring the metropolitan area (usually) grew for an assortment of personal, cultural, economic and political reasons, most having little to do with the best use of tax dollars or the reality of services." (37) Not surprisingly, these fragmented governments struggle to provide even the most basic services. Larger cities are experiencing the same problems.

It is these regional economies that foster quality places vibrant downtowns, attractive town centers and historic, older suburbs that feed the development and acquisition of human and financial capital, and contribute to resource-efficient, sustainable growth. Sadly, the Greater San Diego region desperately needs such consolidation not just of the city and county, but all the municipalities.

What must we do to develop the "smart community?" We must organize ourselves to create a vision, a plan and a "Collaboratory" a new decision-making mechanism for the digital age. At the heart of the Collaboratory is recognition of the importance of cooperation, collaboration and consensus decision-making.

As cities awaken to the challenges of fast becoming the places where creativity gives birth, only art and cultural activities can strengthen a community, particularly when they reveal and celebrate its character and identity. For example, by creating places where people can congregate and watch a performance, purchase a coffee, look at public art and simply enjoy another's company. This is known as creative placemaking, a vital spark that brings a neighborhood to life, making it a place where things are happening, and people want to be. Creative placemaking can mean renovating a historic theater or building affordable live-work space for artists. It can entail transforming a weedy lot into a lively gathering place or an abandoned church into a community exhibition space. It can be organizing a dance festival or providing opportunities for youth to discover their creative power.

While we are only beginning to see technology explode.

REFERENCES

Akhtar, Norman, and Hasley, Smart Cities Face Challenges and Opportunities, Computer Weekly, July 25, 2018.

Anderson, James and Jolin, Michele, Rebuilding Trust in Government, US News and World Report Aug. 5, 2019.

Decant, Skip, Is Smart City just a Euphemism for plain old Surveillance? Government Technology Magazine, February 4, 2021.

Eger, John, California's Effort Promoting Art and Culture Districts, Huffington Post, December 16, 20-17.

Eger, John, Creative Age Cities, Huff Post, 2015.

Florida, Richard, The Economic Power of Art and Culture, Bloomberg CityLab Newsletter, March 19, 2019.

Frost-Kumpt, Hillary-Ann, Art Districts, Americans for the Arts, December 31, 1991.

Goldsmith, Stephen and Crawford, Susan, The Responsive City: Engaging Communities Through Data-Smart Governance 1st Edition, John Wiley and Sons, 2014.

Grantmakers in the Arts, Annual Report, 2019.

Green, Jared, Why Public Art Is Important, The Dirt, October 15, 2012.

Ibid

Ibid

Ibid

Ibid

Ibid

Ibid

Ibid

Ibid

Ibid

Ibid

Ibid

Jankowski, Simona, IoT as the 3rd Wave of the Internet, Goldman Sachs, September 4, 2014.

Katz, Bruce, The Metropolitan Revolution, Governing, June 2013.

Kocieniewski, David, A Wealth of Municipalities, and an Era of Hard Times, The New York Times, May 29, 2009.

Lindskog, Helena, Smart Community Initiatives, University of Linkoping, January 2004.

Ohmae, Kenichi, The Rise of the Region State, Foreign Affairs, 1993.

Poole, Steven, The Truth about Smart Cities: In the end they will destroy democracy, The Guardian, December 17, 2014.

President Obama Addresses Joint Session of Congress, The Washington Post, February 24, 2009.

Public Arts Advisory Council, Why Public Art Matters: Green Paper, American for the Arts ,2014.

Roberts, Ron, we need cooperation, not a merger, San Diego Tribune, March 1, 2009.

Schupback, Jason, Defining Creative Placemaking, National Endowment for the Arts, 2002.

SmartCityPHL Roadmap City of Philadelphia.

Smith, Joshua Emerson, As San Diego increases use of street light cameras ACLU raises Surveillance Concerns, LA Times, August 5, 2019.
Snook, Alastair, what is the purpose of public art? BBC newsletter, October 21, 2014.
Spivak, Miranda. New York Times, January 20, 2021.
Thompson Reuters, Building the Communities of Tomorrow, eRepublic, 2018
Webb, Willington, This Is the Century of Cities, POLIS, (before the US Conference of Mayors) 2009.

Chapter 8

The Technology Explosion

We talk about our IPhone, our I Pad, our smart watches and now sites like Facebook, Twitter, Integra and Zoom dominate our online world. The world has quickly embraced technology, but I am not clear we fully comprehend how it has changed how we work, how we live and play or its revolutionary impact on society and the world economy.

The World of Digits

It began with digitization and the evolution of the computer. The digital world was made possible when Bell Labs, the R&D arm of AT&T, found a way to turn language and communications of any kind, into binary digits. Before that everything was measured in waveforms: amplitude, time and cycles per second. (1)

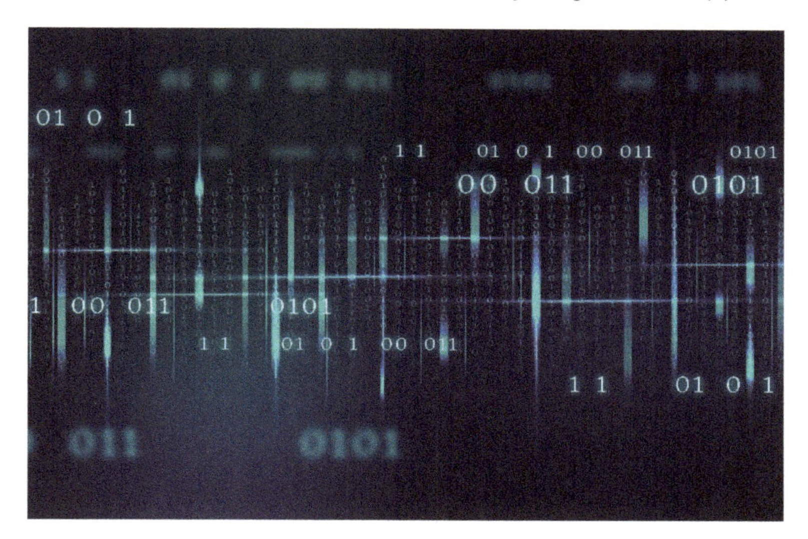

Then the Defense Department (DOD) and needed a communication method that made it extremely difficult to eavesdrop or hack its messaging. The program was called DARAPNET for the network created by The Advanced Research Projects Agency of the DOD. Almost 20 years later the DARPANET became the Internet followed by the Worldwide Web, and the dot.com explosion took flight. (2)

With satellites at 22,500 miles in space, electronic watches getting smarter every day, microwave and cable and Wi-Fi systems dotting the landscape, everything is tied to everything else. The global village is here. It's not on all fours with the village concept but still an indication of the technological advances that have happened in the last two decades.

Compared to what we experienced 100 or so year ago, what has occurred is beyond our comprehension. For example, historians say thousands attended Lincoln's historic address in 1839 at Gettysburg. But because of his high-pitched squeaky voice, and, of course, no sound systems, only about 200 heard it. (3)

Fast forward and 100 years ago, JFK was shot in Dallas, Texas and billions saw, read or heard about the assassination within 24 hours. When the twin towers fell, most people in world saw it instantly. The point is that we have made such progress that we have shrunk time and space and, in the process, as Thomas Friedman acknowledged, shrunk the world.

Gaps are Closing Quickly

The Internet of Things (IoT), also sometimes referred to as the Internet of Everything (IoE), consists of all the web-enabled devices that collect, send and act on data they acquire from their surrounding environments using embedded sensors, processors and communication hardware. Now the idea of connecting almost anything, of any size or complexity is a reality.

In December 1995, MIT's Nicholas Negroponte and Neil Gershenfeld write in "Wired Magazine": "For hardware and software to comfortably follow you around, they must merge into software. The difference in time between loony ideas and shipped products is shrinking so fast that it's now, oh, about a week." (4). Gershenfeld also writes in "When Things Start to Think": "Beyond seeking to make computers ubiquitous, we should try to make them unobtrusive. The real promise of connecting computers is to free people, by embedding the means to solve problems in the things around them." (5)

Most innovations that followed are advances that get information or data from A to B cheaper, faster, better. For example, when Fiber optics was used to send signals over a wire or twisted cable, dramatically expanded transmission of TV signal, data and voice.

The same thing happened with cable itself, so much so we measure cable capacity in terms of bandwidth the amount of data that can be sent from one point to another in a certain period of time. It is measured as a bit rate expressed in bits per second (bits/s) or multiples of it (Kbit/s Mbit/s etc.).

Broadband and Increasing Bandwidth

Broadband is simply very wide bandwidth transmission, which can vary, county to county. The size of a nation has a lot to do with the cost of implementing Internet infrastructure. Having to run fiber optic backhauls is expensive, so running fiber optics across a smaller and dense population in a nation like Korea is considerably cheaper than wiring up the United States so your investment goes further. The US has bandwidth operating at lower speeds than Norway or South Korea.

Many people ask what the limit of broadband is? Frankly, it's limitless. It is the infrastructure of the future. That is why broadband is as important as the waterways, railroads and interstate highways of an earlier era. Cities of the future will decidedly have 24/7 broadband telecommunications in place, wired and wireless infrastructures connecting every home. School and office and through the World Wide Web to every organization or institution worldwide.

In late 2019, Australian researchers found a way to enhance delivery at even faster speed allowing for example, 100 high-definition movies to be downloaded in less than a second. (6)

In less than a decade, the great global network of computer networks called the Internet has blossomed from an arcane tool used by academics and government researchers into a worldwide mass communications medium. It has become the lead carrier of all communications and financial transactions affecting life and work in the 21st Century.

Internet 2?

Grid computing experiments tying hundreds of computers together enabling large scale projects operating at 100 times the speed of ordinary sites, are everywhere. And with cloud computing providers can store data normally at an individual server, and more work can be done, more cheaply at the individual site or promises to in the cloud. 5G, the next generation of Wi-Fi, will operate at 100 times faster and more cheaply too, opening the way for all types of IoE systems.

In 1965, Gordon Moore, co-founder of Fairchild Semiconductor and the CEO of Intel, predicted a doubling of the speed of computing every other year. That phenomena are called "Moore's Law" is a computing term, which he originated. The simplified version of this law states that processor speeds, or overall processing power for not only for computers, but now almost every information system or device will double every two years.

We are witnessing in every sector Moore's law at work and with it more applications that vastly increase productivity. This is good news for the economy but not always for job seekers. Now robots coupled with artificial intelligence enabled by Big Data, are the real threats to jobs.

Global Networks

Many nations are unhappy with the Internet as it exists today and are looking for alternatives, from extranets to intranets to special controls of the existing net. Balkanization or as it is also called, splinternet, is a reality.

Most nations don't like the U.S. influence, real or perceived and believe there are too many sites they would rather their citizens not have access to or that they currently block, for example, porn sites. Many are concerned with the fact, as Edward Snowden revealed, that the U.S. and perhaps others are spying on them. The U.S.s NSA allegedly listened to cell phone calls of Germanys earlier Prime Minister, Angela Merkel. (7)

If nations create their own versions of the net it will be difficult to conduct global business. Corporations, large and small, are likely to find ecommerce more difficult. Many corporations also would rather have offerings with grater bandwidth or transmission at greater speeds.

While nations have the authority to create their own versions of the net, some providers are ignoring use of the net as it is offered, and without objection (so far) are creating virtual private networks, offering more bandwidth at speeds up to 100 Gigabytes, using cloud computers, at very competitive prices. One such provider such as PacketFabric (8) offers network capabilities on a monthly basis or as needed. Many others like Megaport, Aikira and Aviatatre are beginning to offer similar competitive services.

Megaport, and these companies clearly are offering, as Megaport describes it, the "Next generation of Connectivity". Explaining this new network capability, PacketFabric says "today's network infrastructure is much like the electrical grid; it's lagged behind the rest of world and hasn't kept pace with advancements in technology. Procuring network services to provision simple services to move data from one point to another takes months of work and manpower. We're talking months of preparation and operational cost all to move data from Los Angeles to New York in 65 milliseconds." (9) All these companies benefit for the rapidly emerging concept of "Cloud Computing."

Now the Robots

Most robots are nothing more than computers with the ability to crunch reams of data, process that data against a formula and deliver results. They are not thinking machines created by more complex software. But with the accumulation of all kinds of online searches, web site browsing, online purchases, likes and dislikes the Robot can know more about you than your mother does.

A recent study by the Mckinsey Global Institute (10) forecasts up to 800 million workers worldwide could lose their jobs to automation by 2030. Industrial machine operators, administrators, and service workers will be the first to take a hit. Meanwhile, poorer countries with lower investment in tech are less likely to feel the pinch.

This development shows that while humans are handing over a larger share of labor hours to their robot counterparts, the future isn't all bleak. Although 75 million jobs could be displaced by the coming shift in labor, there will be 133 million new jobs created as well. While certain jobs are becoming redundant, human skills remain in demand in other areas.

Robots are less likely to take over roles dependent on human interaction like doctors and teachers. Workers in specialized roles, such as plumbing and health care work, can breathe easy too. Jobs in manufacturing, transport, and administration may decrease. But a potential rise in health, science, tech, and hospitality jobs is likely to offset this trend.

So the real question is, will robots replace your job or make room for you to pursue a new career? Probably not as much as you might think but more than you might imagine. The McKinsey study shows that while humans are handing over a larger share of labor hours to their robot counterparts and although 75 million jobs could be displaced by the coming shift in labor, there will be 133 million new jobs created. While certain jobs are becoming redundant, human skills remain in demand in other areas. (11)

According to the American Association for the Advancement of Science, we are in the midst of a revolution. They say, "In field after field, the ability to collect data has exploded—in biology, with its burgeoning databases of genomes and proteins; in astronomy, with the petabytes flowing from sky surveys; in social science, tapping millions of posts and tweets that ricochet around the internet. The flood of data can overwhelm human insight and analysis, but the computing advances that helped deliver it have also conjured powerful new tools for making sense of it all." (12) As the explosion continues, we are best in a position to shape it.

REFERENCES

Andrews, Evans, Who Invented the Internet, History, December 18, 2013.

Appenzeller, Tim, The AI revolution in science, AAAS, July 7, 2017

Gershenfeld, Neil, When Things Start to Think, 1st Edition, Henry Holt, 1999.

Gilmore, Jezzibell How to be a disrupter, Podcast, July 29, 2019.

Gramenz, Jack, Australian researchers shattered a previous record for the fastest internet speed achieved, and it could fix some of the NBN's biggest problems, News, April 21, 2020.

Ibid

Manyika, James. et al. Jobs lost, jobs gained, McKinsey Global Institute, November 28, 2017.

M. M. Irvine, "Early digital computers at Bell Telephone Laboratories," in IEEE Annals of the History of Computing, vol. 23, no. 3, pp. 22-42, July-Sept. 2001.

Negroponte, Nicolas and Gershenfeld Neil, Wearable Computing. WIRED, December 1995.

PacketFabric names former Cisco executive Ward as its new CEO, FIBERGUIDE, April 22, 2020.

Reuters Staff, U.S. Spy Agency tapped German Chancery, Reuters, July 9, 2016.

Wills, Garry, Lincoln at Gettysburg: The Words that Remade America (Simon & Schuster Lincoln Library), 1992.

Chapter 9

The Man/Machine Interface in the Post Pandemic World

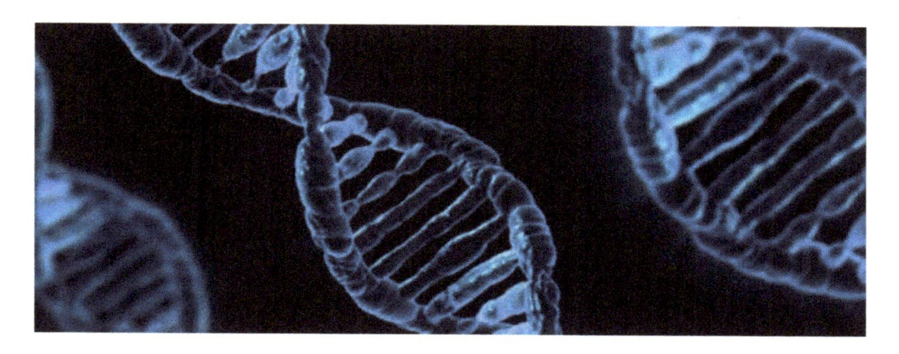

No one really knows how industry will replace humans with machines; what the cost will be to those who employ them, and which workers will be let go. It is also not clear of if they will be retrained, if retaining is possible or what we will do in the way of compensation if we cannot.

We are barreling into unknown territory. Nor do the experts know how the potentially paralyzing effects robotics will have on the workplace or how we will be able to use technology effectively in our cities that maximize affordability, accessibility and while minimizing the adverse effects of using the Internet of Everything".

The urgent need for people with new thinking skills to capture the new jobs is escalating much beyond our comprehension. There seems no doubt that jobs will be lost, some new ones will be created but the new jobs, jobs we are just beginning to define-like "Data Scientist" or "Blockchain Developer" are going to require advance learning, new skills, a different understanding of the world and the world economy. Daphne Koller, President and co-founder of the online University, brand new source of jobs brought about by the flood of users on Facebook, twitter, email, and other sites gathering consumer habits.

According to the Wall Street Journal, "While a six-figure starting salary might be common for someone coming straight out of a doctoral program, data scientists with just two years' experience can earn between $200,000 and $300,000 a year, according to recruiters."

Vinton Cerf, often called "the Father of the Internet" put it this way:

> "What should be fairly obvious, on reflection, is that new jobs created by innovation often require new skills and some displaced workers may not be able to

> learn them. Even when there is a net increase in jobs resulting from innovation (think of the invention of the integrated circuit, the World Wide Web, YouTube), not everyone displaced will find new work unless or until they are able to learn new skills or apply new knowledge."

As the Conference Board, a New York based 'Think tank" for major corporate interests, and Americans for the Arts discovered after surveying 155 U.S. business executives (employers) and 89 school superintendents and school leaders, the number one skill was "creativity". (1)

"Innovation is crucial to competition, and creativity is integral to innovation", they reported. "Overwhelmingly, both the superintendents who educate future workers and the employers who hire them agree that creativity is increasingly important in U.S. workplaces, yet there is a gap between understanding this truth and putting it into meaningful practice. Among the key findings of this research:

- 85 percent of employers concerned with hiring creative people say they can't find the applicants they seek.

- Employers concerned with hiring creative people rarely use profile tests to assess the creative skills of potential employees. Instead, they rely on face-to-face interviews.

- While 97 percent of employers say creativity is of increasing importance, only 72 percent say that hiring creative people is a primary concern." (2)

But why creatives, people with those creative skills, will be in such demand and how someone becomes creative are still unknown. Being creative is very much in discussion.

A Whole New Mind

The key may be staying well ahead of the progress of the Robot; that is what Daniel Pink, author of *A Whole New Mind* (3), seems to be saying; that unless we develop our skills on the right side of our brain to be more empathetic, intuitive, to see forests and trees and thus be more creative, we will not outpace the robot that will eventually acquire those skills. Most believe they will not.

If anything is repetitive however, or merely requires software that enables speed, logic, or mathematical excellence, machines will always be better, faster and cheaper. "The future," Pinks says, "Belongs to a very different kind of person with a very different kind of mind-creators and empathizers, pattern recognizers and meaning makers. These people-artists, inventors, designers, story tellers, caregivers, big picture thinkers-will now reap society's rewards and share its greatest joys." (4)

The right brain, Pink and other who have pondered the two hemispheres of the brain, believe that it is the right Brain that makes use "human." And the human will, perhaps always, be in charge. This should mean the robot, productive as it is, and the human or human brain possessing, as Noble Prize winner James Watson who discovered DNA, said: "Is the most complex thing we have yet discovered in our universe." (5) Man and machine can coexist. There are just too many jobs that machine can absorb but more that the machines will create that only humans can do.

Productivity = Robots

The increase in productivity achieved by technology is too good to ignore. The bad news is that as Pew Research Center believes, we will see tremendous job loss and with "vast increases in income inequality, masses of people who are effectively unemployable, and breakdowns in social order." (6) According to a recent study by the McKinsey Global Institute, 800 million workers worldwide "could lose their jobs to automation by 2030." (7)

Supporting those dire assessments, the World Economic Forum (WEF) Research report out of Davos, projects that by 2020, 7.1 million jobs are expected to be lost, and two million gained, with a net impact of five million jobs lost in the next half-decade." (8)

The future isn't all-bleak. Although 75 million jobs could be displaced by the coming shift in labor, there will be 133 million new jobs created as well. While certain jobs are becoming redundant, human skills remain in demand in other areas.

The WEF report asks us to "think of Artificial Intelligence (AI) as a "black box": we put knowledge in the box; then a little bit more knowledge; and a little bit more". (9) But in the end, we can't currently get anything out of the box except for what we've put into it. So, "thinking outside the box" will take on a whole new meaning in the age of AI. For us, there will be two key developments:

- first, there are still limits constraining AI's progress. But in the future, machines will increasingly learn independently. They'll become capable of "thinking outside of the box".

- Second, it is understandable that the increasing involvement of AI in our lives may arouse fears and anxieties and we must take these fears seriously. But we always have to keep in mind that just as humans have guided and driven the development of AI to this day, they'll continue to lead in the future." (10)

Catesby Holmes, Global Affairs Editor of the "Conversation," an academic journal, writes in the WEF report that "today, robots and smart systems are servants that work in the background, vacuuming carpets or turning lights on and off. Or they're machines that have taken over repetitive human jobs from assembly-line workers and bank tellers. But the technologies are getting good enough that machines will be able work alongside people as teammates much as human-dog teams handle tasks like hunting and bomb detection." (11)

Holms says that robots will continue to be of service to humans, particularly the military, serving as drones, or bomb disposers, but that they "will soon start working in fields as diverse as health care, agriculture, transportation, manufacturing and space exploration." (12)

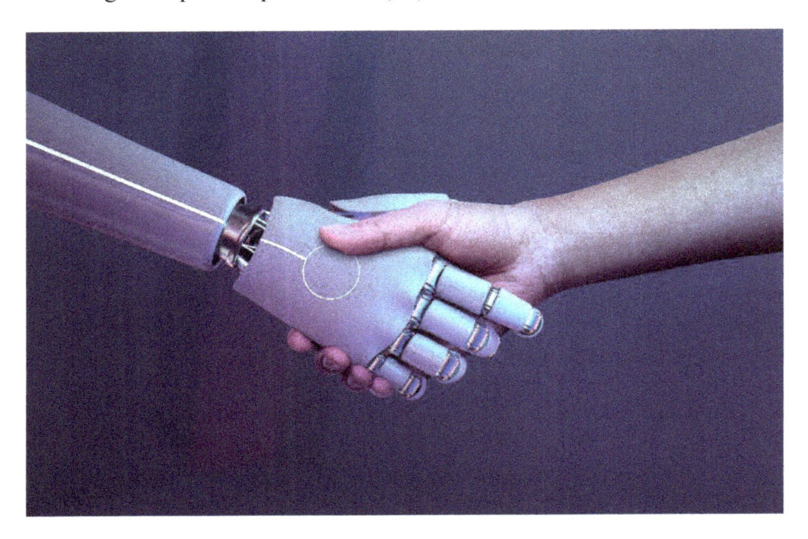

Collaboration Key

Writing for "Wired Magazine", Matt Simon says, "The promise of robots in the age of the new coronavirus is that, theoretically, they're the ideal medical professionals. They don't get sick, they don't need breaks, and they can do menial tasks like delivering supplies. All of these would free up real doctors and nurses to tend to patients. But in 2020, medical robots are still a bit crude. That is changing fast. Globalization 3.0, and the sophisticated developments of Artificial Intelligence aren't threatening enough, the virus pandemic has accelerated the demand for more automation." (13)

There are many developers who want robots to do everything with human assistance. Take "driverless cars for example. But as John Markoff who interviewed Ben Schneiderman, a University of Maryland computer scientist, who believes that such efforts are headed in the "wrong direction". "It is a warning that's likely to gain more urgency when the world's economies eventually emerge from the devastation of the coronavirus pandemic and millions who have lost their jobs try to return to work. A growing number of them will find they are competing with or working side by side with machines." (14)

Schneiderman, says, "Instead of trying to create autonomous robots, he said, designers should focus on a new mantra, designing computerized machines that are "reliable, safe and trustworthy." (15)

Matt Toussaint, Ph.D. Senior Vice President, Product & Content Operations, of a division of the American Chemical Society, believes that collaboration is the only way forward. He writes, "This collaborative "give-and-take" is the cornerstone of the human-machine relationship today. Machines enable humans to process large volumes of information faster and solve more challenging problems by finding patterns in that data. Likewise, humans enable the technology to evolve and deliver the best possible results. (16) No doubt this what is the answer to our concerns about being replaced?

Tusaant also believes that "In the future, we will confront problems that don't even exist today. Our quest for new discoveries and innovation will become increasingly complex and the amount of data available will be unimaginable." (17)

Without a doubt, we will need machines more than ever to help us navigate and make sense of it. But ultimately, humans will continue to be essential to the process as well, setting new constructs that enable greater machine learning and applying machine-gleaned insights to drive new discoveries.

The Public Policy Response

Andrew Yang, Presidential candidate, talked a great deal about the concept of a "Universal Basic Income (UBI)." The idea of getting free money certainly "caught people's attention, but there seemed little understanding of why the concept needs to be discussed.

The idea of the UBI is simple. Give anyone, usually at or near the poverty level, a cash infusion weekly, maybe monthly to secure their basic needs for groceries and other necessities of life, although the payments can be spent on almost anything.

The concept actually goes back to the early 20th century. More recently it has been tried in Switzerland, and parts of Canada but has been discussed in the US and in Europe more recently as a way to provide support for people who lose their job or are experiencing a severe reduced work schedule.

Technology advances are expected to cause adoption of a Basic Income. But this in itself is unlikely to satisfy the needs of the worker who has been replaced by a robot.

Darrell West of the Brookings Institute thinks that "there have to be ways for people to get health care, pension, disability, and income supplements outside of full-time employment." (18) A basic Income, he argues, is not going to be enough. Government should look at "revamping the earned income tax credit, providing activity accounts for lifetime education and retraining, expanding corporate profit-sharing, and providing benefit credits for worthy volunteerism". (19)

Reforming, the school curriculum for the new jobs and encouraging continuing education and access to arts and culture, especially for adults so they can expand their horizons throughout their lives. is mandatory. All this to insure in the long term, that people have the skills needed to work in the new robotic economy. Many economists also believe that the UBI somehow encourage people to be a volunteer and seek out ways to help existing non-profits. In fact, West says, "A variety of survey evidence demonstrates that young people are particularly interested in volunteerism. In general, they have different attitudes towards work and leisure time, and many say they want time to pursue outside activities. For example, a survey of American students found that they want "a job that focuses on helping others and improving society." In addition, they value quality of life considerations, not just financial well-being." (20)

References

Holmes, Catesby, World Economic Forum, 2018.
I bid
I bid
I bid
I bid
I bid
I bid
I bid
I bid
I bid
Markoff, John, A Case for Cooperation Between Machines and Humans. New York Times, May 21, 2020.
McKinsey Global Institute, Jobs lost, jobs gained: The Future of Work.
Pink, (10), Pink, Daniel, A Whole New Mind, Riverhead Books, 2005.
Schaeffer, Catherine, PEW Research Center, February 7. 2020.
Simon, Matt, Spot the Coronavirus Doctor Robot Dog Will See You Now, WIRED Magazine, April 24. 2020.
Toussant, Matt, An evolving partnership: The future of man and machine, CAS Blog, September 7,2018.
West, Darrell, what happens if a Robot takes my job? Brookings, October 2005.
Woock, Christopher, Lichtenberg, Jim, Wright, Mary, Ready to Innovate, The Conference Board, November 2017.
World Economic Forum, WEF Research Report, 2018.

Chapter 10

Conclusion

In his book of Steve Jobs, Walter Issacson discovered that in all Apple's products "technology would be married to great design, elegance, human touches and even romance." (1) Issacson also added, "The creativity that can occur when a feel for both the humanities and the sciences combine in one strong personality was the topic that most interested me in my biographies of Franklin and Einstein, and I believe will be key to creating innovative economies in the 21st century." (1)

If we are keen to do something yet troubled there are few guides or policies to follow, we cannot be deterred. It may seem exaggerated to say this but if we fail to change or to adapt to the new innovation economy and meet the challenges facing us, then we will see more jobs lost, lost forever; the new jobs out of our reach, and the slow but inevitable decline of the species. As a whole new economy based upon creativity and innovation emerges, the importance of reinventing business strategies, corporations, communities, and importantly, our schools, is critical.

Jobless Recovery

Many studies are now in agreement that jobs will be lost, perhaps automated, and that the new jobs created, only if you have the new thinking skills of the most creative and innovation worker. But is the workforce going to have such skills?

If we are to prepare the current and future generations to enter the creative and innovation workforce, we need to redesign our K-12 and college curricula if we are to survive, let alone succeed, in this new global economy.

Many experts tell us not to worry. We have faced change-much as a result of advances in technology- before, and as many new jobs created as are lost. Maybe, but will we have acquired the new thinking skills to fill those new jobs and do so in a timely fashion? Probably not.

A New Urgency

As I mentioned in my introduction, former Google CEO Eric Schmidt, said there was a sort of silver lining for tech companies as investment in technology was likely to dramatically increase; and as month's experts, critics and journalists all are saying the same thing: After the Pandemic life will never be the same. And as already noted by Alexandra Ossola: "We have fields like telemedicine that in a matter of weeks jumped ahead to where we thought we would be in 10 years. Robotics is being deployed into roles like policing curfews, to cleaning subways and hospitals, to delivering groceries." (3)

This is probably more important than anything you or I have said or read or heard elsewhere. Maybe the Coronavirus, if we can find anything good to say, is we as a people seem to be more understanding of each other. We are aware of the importance of cooperation; the urgent need to work together. We do know we simply cannot get anything we think we might accomplish done alone.

Retraining will be necessary, so too basic changes in the curriculum. An art-infused, integrated curriculum, is a logical answer to nurturing both hemispheres of the brain and producing the kinds of skills we desperately need to compete. While there are many things, we must do to reinvent our schools for the new economy, we must take a fresh look at the critical role of the arts in transforming our curriculum. The Basic Uniform Income (BUI), awarding monetary grants on a monthly basis to those out of work is also being discussed.

Recent reports say Business and academia are not aligned, saying the educators still don't get IoT/IoE. So too, cities, communities, everywhere. Both need to understand the challenges we face in the new economy, an economy where AI, Robots, algorithms and IoT overwhelm us all.

This is more than rhetoric. This is a warning that we cannot avoid the fact that the jobs that exist today will be gone over the next several years. The new jobs-the jobs that come with the new creative and innovation economy require new thinking skills. We don't know yet what the new jobs are or frankly the new skills we know will be needed, we do know that robots cannot do those jobs that will be available; they are jobs that only humans can do. They will require empathy, creativity and an ability to work with robots themselves.

This awareness of the combined power of the arts and sciences, between art and technology, is vital. It demands that we renew and reinvent ourselves to better understand the challenges and our communities atrophy and die, and with-it great experiments in freedom and the human spisrit, of a global economy, the power of the arts and sciences, and the esthetic makeup of our cities and communities where we live and work. There is no alternative.

We are barreling into unknown territory. No one really knows how we will be able to use technology in our cities that maximize affordability, accessibility and usefulness and minimize the adverse effect using the Internet of Everything" (IoT) or the paralyzing effects robotics will have on the workplace.

Nor does anyone really know how industry will replace humans with machines; what the cost will be to those who employ them, and which workers will be let go. It is also not clear of if industry is going to use robots. No doubt. "Meeting in Davos last year, Jena McGregor of the "Washington Post", reported that those attending were enthusiastic about what they called the "Fourth Industrial Revolution (the) term for the accelerating pace of technological changes, especially those that are "blurring the lines between the physical, digital and biological spheres" the combination of things like artificial intelligence, robotics, nanotechnology and 3-D printing."(4)

The Demand for Smart Machines

The increase in productivity is too good to ignore. The bad news is that as Pew Research Center believes, we will see tremendous job loss with "vast increases in income inequality, masses of people who are effectively unemployable, and breakdowns in social order." (5). Industrial machine operators, administrators, and service workers will be the first to take a hit. Meanwhile, poorer countries with lower investment in tech are less likely to feel the pinch.

The WEF, McKinsey and other studies all show while humans are handing over a larger share of labor hours to their robot counterparts, the future isn't all-bleak. Although 75 million jobs could be displaced by the coming shift in labor, there will be 133 million new jobs created as well. While certain jobs are becoming redundant, human skills remain in demand in other areas.

So the real question is, will robots replace your job, or make room for you to pursue a new career? It really depend on your current skill set but probably not as much as you might think but more than you might imagine.

Forty seven percent of all jobs are likely to face automation over the next 20 years. However, the same study reveals 53% of jobs are unlikely to be affected at all. Robots are less likely to take over roles dependent on human interaction like doctors and teachers. Workers in specialized roles, such as plumbing and care work, can breathe easy too. Jobs in manufacturing, transport, and administration may decrease. But a potential rise in health, science, tech, and hospitality jobs is likely to offset this trend.

So the real question is, will robots replace your job, or make room for you to pursue a new career? It really depend on your current skill set but probably not as much as you might think but more than you might imagine.

Nonetheless, according to the American Association for the Advancement of Science, we are in the midst of a revolution. They argue that: "In field after field, the ability to collect data has exploded—in biology, with its burgeoning databases of genomes and proteins; in astronomy, with the petabytes flowing from sky surveys; in social science, tapping millions of posts and tweets that ricochet around the internet. The flood of data can overwhelm human insight and analysis, but the computing advances that helped deliver it have also conjured powerful new tools for making sense of it all." (7)

An exhaustive report by David Berrby in a recent issue of National Geographic, a gloomy picture is drawn because as Berrbey points out, "Robots can be programmed or trained to do a well-defined task—dig a foundation, harvest lettuce—better or at least more consistently than humans can. But none can equal the human mind's ability to do a lot of different tasks, especially unexpected ones. None has yet mastered common sense." (9)

Art May Save the World

Last thoughts.

In 2011 declaring October as National Art and Humanities Month, President Obama made the observation. "Educators across our country are opening young minds, fostering innovation, and developing imaginations through arts education. Through their work, they are empowering our Nation's students with the ability to meet the challenges of a global marketplace. It is a well-rounded education for our children that will fuel our efforts to lead in a new economy where critical and creative thinking will be the keys to success."

More and more people in high places seem to be saying the right thing. Last April, Arne Duncan, U.S. Secretary of Education, said: "The Arts can no longer be treated as a frill. Arts education is essential to stimulating the creativity and innovation that will prove critical for young Americans competing in a global economy." But we have seen too little in the way of action.

Is this because the administration really doesn't believe what they say about the arts? Because Washington, D.C. can't get anything done? Or because the benefits are still not obvious to most politicians.

I really don't know, but I do know that not every parent knows his or her children may not get a job. Most politicians seem unwilling to change STEM to STEAM or correct the mistakes of No Child Left Behind. Education executives seem too busy fighting budget woes and teachers are unwilling to take the risk of collaborating with artists.

Yet, recognizing the vital role of the arts in the new economy may be the most important aspect of a 21st century education. Globalization 3.0 is here. Outsourcing jobs, and off-shoring whole divisions of companies are commonplace. We are currently suffering what economists are euphemistically calling a "jobless recovery," and our communities and our schools are facing challenges not well understood by politicians, policymakers, or parents.

Twenty years ago, it was fashionable to blame foreign competition and cheap labor markets abroad for the loss of manufacturing jobs in the United States, but the pain of the loss was softened by the emergence of a new services industry. Now, it is the service sector jobs that are being lost. This shift of high-tech service jobs will be a permanent feature of economic life in the 21st century.

Understandably there is concern.

The New York Times, writing about the PISA tests, interviewed U.S. Secretary of Education Arne Duncan, who, despite his appreciation for the arts, said: "We have to see this as a wake-up call. I know skeptics will want to argue with the results, but we consider them to be accurate and reliable, and we have to see them as a challenge to get better," he added. "The United States came in 23rd or 24th in most subjects. We can quibble, or we can face the brutal truth that we're being out-educated."

Well, forget PISA, and for that matter the platitudes coming out of Washington, too.

Artists, art educators, and art and cultural organizations across the country know well that students who learn through the arts or have the personal experience of studying art, are changed forever. He or she is a different person. They see the world differently. To paraphrase Robert Kennedy, they see the world the way it could be not just the way it is.

"Most of us appreciate the intrinsic benefits of the arts their beauty and vision; how they inspire, soothe, provoke, and connect us," as Bob Lynch of Americans for the Arts once noted. But important today, as Sandra Ruppert, President of Art Education Partnership observed, is this "Arts learning experiences play a vital role in developing students' capacities for critical thinking, creativity, imagination, and innovation. These capacities are increasingly recognized as core skills and competencies all students need as part of a high-quality and complete 21st-century education."

Steve Jobs once said, "People with passion can change the world for the better." Only artists have the passion, and the knowledge to change the world to change so many things with what's wrong today.

Five years ago at Davos , Olafur Eliasson an artist known worldwide and who received a Crystal Award, honoring "artists whose important contributions are improving the state of the world and who best represent the 'spirit of Davos" was asked to speak about the unique role of art in society. He said: "I am convinced that by bringing us together to share and discuss, a work of art can make us more tolerant of difference and of one another. The encounter with art and with others over art can help us identify with one another, expand our notions of we, and show us that individual engagement in the world has actual consequences. That's why I hope that in the future, art will be invited to take part in discussions of social, political, and ecological issues even more than it is currently and that artists will be included when leaders at all levels, from the local to the global, consider solutions to the challenges that face us in the world today." (10)

David Brooks, columnist for the New York Times, talked about the importance of art particularly the humanities, and not just as a tool for enhancing one's economic prowess, but for what make human, what makes us different from computers, and the skills that separate us from one another. Quoting Robert Kennedy after the assassination of Martin Luther King, he recalled the slaying of his own brother and quoted Aeschylus: "In our sleep, pain which cannot forget falls drop by drop upon the heart until, in our own despair, against our will, comes wisdom through the awful grace of God." (11)

I believe, there can be no more distilled expression of a culture than its works of art. In the coming decade, the challenge for humanity will be whether we can come to grips with the idea of a world community, shared governance and the notion that the differences between us the art and culture and wonder and beauty of those differences must be something we can respect, honor, grow to appreciate and welcome. (12)

These expressions of art as a bridge to a multicultural future are yet more reasons, most powerful reasons why marrying art and technology is so important to our shared future.

References

Artificial Intelligence, AAAS, January 25, 2021.
Berreby, David, The Robots Are Here, National Geographic, September 2020.
Brooks, David, New York Times, May 28, 2020.
Eger, John, Art as a Universal Language, Huff Post, January 25, 2011.
Geiger, A. W., How Americans see automation and the workplace, Pew Research Center, April 7, 2010.
Ibid
Ibid
Issacson, Walter, Steve Jobs: The Official Biography, Simon and Shuster, Oct 24, 2011.
Jones, Tim, Can Art Change the World? World Economic Forum, March 28, 2014.
McGregor, Jena, Robots will be all the rage at Davos, this year, Washington Post, January 19, 2010.
Mims, Christopher, As E-Commerce Booms, Robots Pick Up Human Slack, Wall Street Journal, August 8, 2020.
Ossola, Alexandra, Coronavirus is automating the world even faster, Yahoo Sports, May 28, 2020.
Susskind, Daniel, A World Without Work, Henry Holt and Company, 2020.

BIBLIOGRAPHY

Allen, Maury, You Could Look it up, Times Books, January 1, 1979.

"America Competes Act", Conference Report, House of Representatives, 2007.

America Competes Act, United States of America in Congress, Aug 9, 2007.

"Are They Really Ready to Work", The Conference Board, October 2006.

Andrews, Evans, Who Invented the Internet, History, December 18, 2013.

Appenzeller, Tim, The AI revolution in science, AAAS, July 7, 2017.

"Authentic Connections: Interdisciplinary Work in the Arts", Consortium of National Art Education Associations (AATE, MENC, NAEA, NDEO), 2002.

Balcaitis, Ramunas, Design Thinking models. Stanford school, EMPATHIZE@IT EMPATHIZE. DESIGN. BUILD, JUNE 15, 2019.

Barnes, Melody, Reinvesting in Arts Education: Winning America's Future Through Creative Schools, White House, Reinvesting in Arts Education: Winning America's Future Through Creative Schools, May 12, 2011.

Bell, Daniel, The Coming of Post-Industrial Society. New York: Harper Colophon Books, 1974.

Bell, Daniel. The Coming Post-Industrial Society, New York: Basic Books, 1977.

Berreby, David, The Jones, Tim, Can Art Change the World? World Economic Forum, March 28, 2014.

Blueprint for Creative Schools, A Report to State Superintendent of Public Instruction Tom Torlakson 2015, bfcsreport.pdf New Cities Foundation, http://newcities.org/

Brown, Tim, Roberts, Tim, et al, and Change by Design, Revised and Updated: How Design Thinking Transforms Organizations and Inspires Innovation, Harper Collins, 2009.

Bureister, Misti, From Boomers to Bloggers: Success Strategies Across Generations, Synergy Press, January 28, 2008.

Burnaford, Gail, with Brown, Sally, Doherty, James, and McLaughin, James, "Arts Integration: Frameworks, Research, and Practice, Arts Education Partnership, October 29, 2007.

Caldwell, Ellen C., Can Art Help People Develop Empathy? JSTOR Daily, January 16, 2018.

CAPE, Chicago Arts Partnership in Education, January 2011.

Chanda, Nayan, Bound Together: How Traders, Preachers, Adventurers, and Warriors Shaped Globalization, Caravan Books, 2007.

Colvin, Geoff, Humans Are Underrated: What High Achievers Know That Brilliant Machines Never Will p.40, Penguin Random House, 2015.

Consensus Study Report, The Integration of the Humanities and Arts in Higher Education, National Academy of Science, Engineering and Medicine, 2018.

Consensus Study Report, The Integration of the Humanities and Arts in Higher Education, National Academy of Science, Engineering and Medicine, 2019.

Csikszentmihalyi, Mihaly, Creativity: Flow and the Psychology of Discovery and Invention, Harper Perennial, 1996.

Duncan, Arne, The Well-Rounded Curriculum, Department of Education, April 10,2010.

Eger, John, The Smart Communities Guidebook. Report to CALTRANS, SDSU International Center for Communications, 1997.

Eger, John M., The Creative Community. San Diego: California Institute for Smart Communities, SDSU International Center for Communications, 2003.

Eisner, Elliot E., The Arts and the Creation of Mind, Yale University press, 2002.

Florida, Richard. The Rise of the Creative Class. New York: Basic, 2002.

Florida, Richard, The Rise of the Creative Class: and How it's Transforming Work, Leisure, Community and Everyday Life, New York: Basic Books, 2004.

Friedman, Thomas L., The World Is Flat: A Brief History of the Twenty-first Century, New York: Farrar, Straus and Giroux, 2005.

Gardner, Howard, and Frames of Mind: The Theory of Multiple Intelligences Kindle Edition, Basic Books, and November 23, 1983

Gates, Bill, The Speed of Thought: Succeeding in the Digital Revolution, Warner Books, and March 1996.

Gene D. Cohen, The Creative Age: Awakening Human Potential in the Second Half of Life, Avon Books, a division of Harper Collins, 2000.

Gene D. Cohen, The Creative Age: Awakening Human Potential in the Second Avon
Gershenfeld, Neil, When Things Start to Think, 1st Edition, Henry Holt, 1999.
Goldstein, M., Decade of the Brain, US National Library of Medicine, September 1994.
Goldsmith, Stephen and Crawford, Susan, The Responsive City: Engaging Communities Through Data-Smart Governance 1st Edition, John Wiley and Sons, 2014.
Hagel, John, Brown, John Seely, and Davison, Lang, The Power of Pull: How Small Moves, Smartly Made, Can Set Big Things in Motion 1st Trade Paper edition, Basic Books, 2012.
Howkins, John, The Creative Economy: How People Make Money from Ideas, Penguin, 2001.
Issacson, Walter, Steve Jobs the Official Biography, Simon and Shuster, Oct 24, 2011.
"Japanese Tasks in the 1990s," NIRA Research Output, Vol.1, Number1, 1988.
Jones, Robert, Scanland, Kathryn, Gunderson, Steve the Jobs Revolution: Changing How America Works, Copywriters Inc., a division of the Greystone Group, Inc., 2005.
Kathryn, Gunderson, Steve, The Jobs Revolution: Changing How America Works, Copywriters Inc., a division of the Greystone Group, Inc., 2005.
Krippendorff, Kaihen, The De
Larson, Gary O., American Canvas, National Endowment for the Arts, January 1, 1997.
Markham, Art, 3 Ways to Train Yourself to be More Creative, FAST Company, April 12. 2015.
McCabe, David, Bill Clinton's Telecom law: Twenty years later, The Hill, February 7, 2016.
McGilchrist, Iain, "The Battle if the Brain", The Wall Street Journal, January 2, 2010.
M. M. Irvine, "Early digital computers at Bell Telephone Laboratories," in IEEE Annals of the History of Computing, vol. 23, no. 3, pp. 22-42, July-Sept. 2001.
Naisbitt, John and Doris, Megatrends, Harper Collins, 1982.
National Association of Manufacturing, Skills Gap Report a Survey of the American Manufacturing Workforce, 2005.
Neuroeducation: Learning, Arts, and the Brain", THE DANA FOUNDATION, January 2011.
Ohmae, Kenichi, The Rise of the Region State, Foreign Affairs, 1993.
Peter F. MacNeilage, Lesley Rogers, and Giorgio Vallortgara, "Evolutionary Origins of Your Right and Left Brain", Scientific American, July 2009.
Porat, Marc. "The Information Economy: Definition and Measurement.
Read, Ash, Everything You Need to Know About Global Marketing Strategy, SUMO, January 21, 2020.
Reetz, David, Bershad, Carolyn, LeViness, Peter, Whitlock, Monica, Annual Survey of Colleges and Universities, on General Education, 2016.
Restak, Richard, M.D., The New Brain, Rodale, 2003.
Revolution in the U.S. Information Infrastructure, National Academy of Engineering, 1995.
Root-Bernstein, Michele & Robert, Sparks of Genius. Boston: Houghton Mifflin, C.P. Snow, Two Cultures, Cambridge University Press.
Root-Bernstein, Robert and Michele, Sparks of Genius: The 13 Thinking Tools of the World's Most Creative People, Houghton Mifflin Company, 1999.
Sawyer, Keith, Zig Zag: The Surprising Path to Greater Creativity, Wiley Imprint, 2003.
Shapiro. Philip, Masser, Ian, Edgington, David, Planning for Cities and regions in Japan, Google books, 1991.
Shlain, Leonard, Art & Physics: Parallel Visions in Space, Time, and Light, Viking, 1977.
Shlain, Leonard. The Alphabet and The Goddess: The Conflict Between Word and Image, Viking, 1998.
Smith, O.C., Little Green Apples: God Really Did Make Them! Simon and Shuster, 2003.
Snyder, Sue, Language, Movement and Music Process Connections' in General Music Today, Spring, 1994.
Staff, Design Thinking What is that? Fast Company, March 20, 2006.
Susskind, Daniel, A World Without Work, Henry Holt and Company, 2020.
Thomas L. Friedman, The World Is Flat: A Brief History of the Twenty-first Century, New York: Farrar, Straus and Giroux, 2005.
Tim Brown, Tim Roberts, et al, Change by Design, Revised and Updated: How Design Thinking Transforms Organizations and Inspires Innovation, Harper Collins, 2009.
Turnali, Kann, What is Design Thinking? Forbes, May 10, 2015.
Wills, Garry, Lincoln at Gettysburg: The Words that Remade America (Simon & Schuster Lincoln Library), 1992.

INDEX

"America is not going to succeed through cheap labor or cheap materials, nor even the free flow of capital or a streamlined industrial base to compete successfully, this country needs creativity, ingenuity, and innovation."

Dana Gioia, Chair, National Endowment for the Arts May 8, 2008.

About the Author

John M. Eger is Emeritus Professor of Communication and Public Policy at SDSU, and Director, The Creative Economy Initiative, a nonprofit educational organization founded to pro mote the concept and facilitate the implementation of "smart and creative communities. Earlier, he was Senior Vice President of the CBS Broadcast Group. From 1973-1976, he was Advisor to Presidents Richard Nixon and Gerald Ford and Director of the White House Office of Telecommunications Policy (OTP). He also served as Chairman of California's first Commission on Information Technology; and San Diego's "City of the Future" Commission.

We are living through the closing chapters of the established and traditional way of life.

We are in the early beginnings of a struggle to remake our civilization. It is not a good time for politicians.

It is a time for prophets and leaders and explorers and pioneers, and for those willing to plant trees for their children to sit under.

Walter Lippman

CPSIA information can be obtained
at www.ICGtesting.com
Printed in the USA
BVHW010319311222
655314BV00003B/136

9 780949 313027